INNOVATION POLICY

WESTERN PROVINCES OF CANADA

ORGANISATION FOR ECONOMIC CO-OPERATION AND DEVELOPMENT

Publié en français sous le titre:

LA POLITIQUE D'INNOVATION
PROVINCES DE L'OUEST DU CANADA

This report is part of the OECD series of Innovation Policy Reviews of individual Member countries.

These reviews have two purposes. First the review enables the country or region concerned to appraise the performance of the institutions and mechanisms which govern or influence the various fields (scientific, technological and industrial, but also economic, educational and social) which contribute to its innovation capacity. The review provides an opportunity to assess the depth of social awareness of the need for, and the causes of, innovation.

Second, the review helps to enrich the pool of available knowledge on the content of innovation policies and their role as an instrument of government. In this way OECD countries can derive lessons which should help them to perfect their own policies. Similarly, through this improved knowledge of the resources deployed by Member countries, the reviews help to strengthen international co-operation.

The reviews are undertaken at the request of governments, which contribute to their cost. A flexible approach is adopted in regard to the focus, the methodology and the presentation of these reviews.

For reviews undertaken at regional level, the process includes firstly provision of background information by the region concerned; such information includes data and analysis of economic and industrial structures, of educational, scientific and technological infrastructures, of financial markets, etc., and documentation on relevant government strategies and policies. The second step is a review mission. The review team is composed of an international group of experts (the "Examiners") with complementary competences in the various aspects of the innovation process and climate, supported by members of the OECD Secretariat who contribute to the analysis and the elaboration of the review report. This team undertakes a series of visits and meetings with government authorities, entrepreneurs, academics, technologists, bankers, representatives of public communities, etc., with a view to grasping the various aspects of the complex reality of innovation and its climate.

Then a report is prepared on the basis of the findings of the review mission and the background documentation. This report is discussed by the OECD Committee for Scientific and Technological Policy (CSTP), meeting in a Special Session.

This general approach is adapted to the specific interests of the region concerned. As regards Western Canada, at the request of the Provincial Governments, the review has considered the four provinces as a unit, in placing their technological development within the national context.

However, each of the four provinces has indicated particular matters to be considered:

- *Alberta* asked that the review be concentrated on agriculture, hydrocarbon related energy industries, and selected components of the electronics industry;
- *British Columbia* asked that the review be focused on electronics and telecommunications and the natural resources based sectors such as forestry and mining;

3

- *Manitoba* asked that particular attention be given to the diffusion of technology throughout the manufacturing sector, to the examination of creative niches and to the adjustment of human resources; and
- *Saskatchewan* asked that the review be focused on the high technology sectors, including comments on policies and programmes.

The review team's visit was planned accordingly, the team spending two days in each province. The mission took place on September 1985, one day being devoted to discussions with Federal Government departments in the federal capital, Ottawa (prior going to the West). A second mission by two members of the review team took place in January 1987. The report was discussed by CSTP in June 1987.

The review team included: Mr. Francis W. Wolek (United States), Professor at the College of Commerce and Finance, Villanova University, formerly Deputy Assistant Secretary for Technology and Innovation, US Department of Commerce (Chairman of the review team); Mr. Francis Bonnet (France), President "Silicon Search" (human resources for high technology industries), formerly Director for training and recruitment, Hewlett Packard Europe; Mr. Michael H. Proctor (New Zealand), international consultant in innovation and technology transfer; Mr. Veikko Vuorikari (Finland), Executive Vice-President, Regional Development Fund of Finland. And two members of the OECD Secretariat: Mr. Jean-Eric Aubert – project co-ordinator – and Mr. Patrick Dubarle, of the Science and Technology Policy Division, Directorate for Science, Technology and Industry.

The review team wishes to express their warmest thanks to the authorities and to the many eminent persons who have been kind enough to contribute to the preparation of this study of innovation policies in Western Canada. It would be impossible to list the names of all the Canadians who have assisted in this exercise.

The review team wishes however to express particular gratitude to the Canadian Western Science Officials Group who were responsible for this study, and more particularly to:

- Its co-ordinator, Dr. Alan Vanterpool, Assistant Deputy Minister, Technology, Research and Telecommunications, Alberta;
- Mr. Phil Gardner, Manager of Science Policy, Industry and Economic Development, British Columbia;
- Dr. Dan Archer, Department of Industry, Trade and Technology, Manitoba; and
- Dr. Alex Guy, Deputy Minister, Science and Technology, Saskatchewan.

The review team would also like to thank the representatives of industry, the universities, and other communities for having enabled it to form a clear picture of the quality and originality of innovation in progress in Western Canada.

Finally the review team would like to express its gratitude to the officials of the Federal Department of Regional Industrial Expansion, the Ministry of State for Science and Technology and the Department of External Affairs who have participated in this study and have been very helpful.

Also Available

INNOVATION POLICY:

SPAIN (December 1987)
(92 87 06 1) ISBN 92-64-13029-2   106 pages

£7.00   US$15.00   F70.00   DM30.00

IRELAND (February 1987)
(92 87 01 1) ISBN 92-64-12918-9   76 pages

£5.00   US$10.00   F50.00   DM22.00

FRANCE (February 1987)
(92 86 06 1) ISBN 92-64-12884-0   296 pages

£16.00   US$32.00   F160.00   DM71.00

REVIEWS OF NATIONAL SCIENCE AND TECHNOLOGY POLICY:

NETHERLANDS (August 1987)
(92 87 03 1) ISBN 92-64-12955-3   142 pages

£9.50   US$20.00   F95.00   DM35.00

SWEDEN (June 1987)
(92 87 04 1) ISBN 92-64-12958-8   112 pages

£6.00   US$13.00   F60.00   DM26.00

FINLAND (April 1987)
(92 87 02 1) ISBN 92-64-12928-6   154 pages

£9.50   US$19.00   F95.00   DM42.00

AUSTRALIA (August 1986)
(92 86 05 1) ISBN 92-64-12851-4   120 pages

£7.50   US$15.00   F75.00   DM33.00

PORTUGAL (June 1986)
(92 86 04 1) ISBN 92-64-12840-9   136 pages

£8.00   US$16.00   F80.00   DM35.00

Prices charged at the OECD Bookshop.

THE OECD CATALOGUE OF PUBLICATIONS and supplements will be sent free of charge on request addressed either to OECD Publications Service, Sales and Distribution Division, 2, rue André-Pascal, 75775 PARIS CEDEX 16, or to the OECD Distributor in your country.

TABLE OF CONTENTS

Part I

REVIEW REPORT

Summary 10

I. Introduction 13
   1. Towards a knowledge-intensive economy 13
   2. Western Canada: unity and diversity 15
   3. Outline of the report 15

II. The West in Transition 17
   1. The economic challenge 17
   2. The high technology-based industry 21

III. Policy Trends 25
   1. Socio-cultural background 25
   2. Provincial policies 26
   3. Federal policies 29
   4. The Economic Regional Development Agreements 30

IV. Human Resources 31
   1. Entrepreneurial resources 31
   2. Scientific and technical manpower 32
   3. Managerial competence 35
   4. Social issues 36

V. Support for New Industry 38
   1. Services for innovators 38
   2. Finance for innovation 42
   3. Research programmes 44
   4. Natural resource economies and innovation programmes 47

VI. Strategies for Growth 51
   1. General considerations 51
   2. The critical mass problem 53
   3. Towards technological identities 55
   4. Regionalisation of innovation policies 59

VII. Policy Recommendations 61
   1. Development of a technological identity 61
   2. Creation of an environment supporting the new industry 62
   3. Enrichment of the human resources 63

4. Need for networking people and organisations 65
5. Extension of inter-provincial co-operation 66
6. Improvement of the relations with the Federal Government 68
7. Specific provincial measures 70

Notes and References 75

Annex: Government Policies 77
       Alberta 77
       British Columbia 79
       Manitoba 80
       Saskatchewan 82
       Federal Government 83

Part II

ACCOUNT OF THE REVIEW MEETING

  I. **Introduction** 88
 II. **General Remarks on the Report** 90
III. **Technological Strategies** 93
IV. **Resources for Innovation and Growth** 97
 V. **Inter-provincial Co-operation** 102

Annex 1: Extracts from the Note Prepared by Mr. Proctor for the Review Meeting 104
Annex 2: List of Participants 106

LIST OF TABLES

 1. Gross domestic product and population 17
 2. Industrial activity by sector 19
 3. Exports by destination 19
 4. Exports by products 20
 5. The micro-electronics-based industry in British Columbia 21
 6. High technology industries in Alberta 22
 7. Advanced technology companies in Saskatchewan 22
 8. Gross expenditures on research and development 23
 9. Funding of regional R&D 23
10. GERD by sector of performance and by province 24
11. Major provincial tax rates 27
12. Education indicators 28
13. Provincial support to university 28
14. Engineering education in Western Canada 33
15. Level of activity in the western provinces of a major US-based computer company . . . 55

LIST OF FIGURES

1. Map of Canada 14
2. Breakdown of GDP – 1984 18
3. British Columbia electronic manufacturers 54

Part I

REVIEW REPORT

SUMMARY

This report reviews policies to promote technology and innovation in the western provinces of Canada, and provides recommendations to policy makers.

The review team, however, would like to emphasize that most of the analyses and conclusions presented in this study are based on information drawn from a short visit which permitted it to observe only a slice of innovative activities in Western Canada. Moreover in dealing with the four provinces, it was difficult to recognise and highlight the uniqueness and aspirations of each in the overall Western Canada region under study. On the other hand, in requesting that they should be approached as a unit, the four provinces have taken an important initiative, of a pilot character in this series of OECD reviews.

The overall scene

Following a brief introduction, Chapter 2 outlines the economic challenge that the western provinces are facing to reduce their dependence on natural resources and agricultural commodities. The Chapter also provides data on the growth of new technology-based industries.

Chapter 3 surveys the policies of the Provincial Governments, as well as the support given by the Federal Government, for stimulating innovation and new technologies in the western economy. Detailed descriptions of these policy initiatives are presented in an Annex to this report.

Human resources

Chapter 4 analyses the development and mobilisation of human resources – the primary resource of a knowledge-intensive economy. One keynote is that the entrepreneurial dynamism of the western provinces is impressive. High quality technologies (the seeds of new industries) continue to be developed in plentiful supply by western entrepreneurs. The ability of Western Canada to turn these innovations into internationally competitive firms would benefit from further enhancement of the pool of executives with international experience in marketing and finance (e.g. recruitment of former residents from foreign, "high-tech" centres, significant expansion of training in entrepreneurship, etc.).

The West, like all of industrial Canada, enjoys a culture which recognises the importance of current sacrifice to build its educational infrastructures (universities, training institutes, community colleges) which produce the knowledge and manpower required by innovative industry. Nevertheless, even larger sacrifices will be necessary to match the potential demand of new enterprises (i.e. significantly greater funding for equipment and internationally experienced faculty). Moreover, better planning of human resources required by new industry is necessary both at national and regional level.

Provincial governments clearly appreciate the process of developing human resources and to ensure an adequate supply of skilled people. However, stronger programmes are needed to involve workers and labour in planning innovations and learning to use new technologies. Special attention to the contributions of women to the West would also be productive. Greater support for their education and encouragement in managerial and technical careers would significantly expand the supply and creative diversity of western competence.

Support for new industry

The West has demonstrated creativity and commitment in the support of innovation (surveyed in Chapter 5), but the infrastructure is still in an early stage of development and there is great scope to expand it (for example, incubator spaces and evaluation centres which are of great help to intending innovators). Measures taken to stimulate venture capital are efficient, but should be expanded with growing experience with the full process of new venture financing and the creation of a Western Stock Exchange is proposed which would provide an "exit mechanism" for venture investors.

It is vital that the universities expand their relations with industry; they should adopt more active policies to encourage industrially-funded R&D and should intensify their co-operation with government and industry, instead of pursuing scientific work in isolation.

Government can only be a catalyst for massive commitments necessary in all sectors of a knowledge-intensive economy and one of the chief findings of this study is the need for significant expansion of the sacrifices and risks taken by the private sector in Western Canada. In addition to private capital and management talent, public spirited leadership is needed. Western Canadian industry, at present, does not provide a competitive level of support for new industry when compared to other OECD countries (support to spin-offs and new ventures, provision of scholarships, testing services to entrepreneurs, etc.).

The difficulty of mobilising support for new industry is particularly apparent in the natural resource sectors, and is explained by the influence of geography and techno-economics on the-day-to-day thinking ("mind sets") of executives leading these industries.

Strategies for growth

The strategies adopted in most provinces (discussed in Chapter 6) emphasize the high technology sectors (telecommunications, CAD/CAM, micro-electronics, biotechnology). More efforts are needed to link high technology to the whole economy and to enrich with new technologies the existing industry, including the natural resource and agriculture sectors. It is important also to pursue active policies for transferring technology from other areas into the provinces, a strategy which might be less costly than the development of indigenous high technology infrastructures.

After discussing the conditions in which the western provinces are reaching critical masses in high technology sectors, and underlying that these sectors are the same targets for most national programmes on innovation, the report recommends that each province develop a technological identity firmly rooted in its strengths (markets, suppliers, infrastructures, talents and leadership). Such a strategic focus would assist the provinces in avoiding dispersion of their resources. Insights are provided on the most promising fields on which each province can focus its efforts.

11

A brief discussion of the regionalisation and decentralisation of innovation and technology programmes within the federal context concludes that giving more responsibility to provincial governments is an efficient way to increase the needed responsiveness of innovation policy to local needs and capabilities. However, steps are warranted to avoid unproductive duplication, within a context of fierce inter-provincial competition for federal funding.

Policy recommendations

The report emphasizes the inter-departmental character of innovation policy and suggests that consideration be given to establishment of a Special Commissioner for Innovation in Provincial Governments.

The report then summarises the recommendations related to the above Chapters:

- Avoidance of fragmentation by focusing on technological identity;
- Development of a more supportive environment for new industry;
- Enrichment of human and knowledge resources.

The report stresses the need for networking people and organisations in appropriate structures (e.g. incubator spaces, evaluation centres, technological associations, biotechnology parks). Managers of these structures should be fully involved in the elaboration of provincial innovation policies, as full members of "guiding councils".

Moreover, the western provinces would benefit from more developed relationships and suggestions are made for co-operative efforts in several areas: e.g. the exchange of experiences, jointly funded research projects in fields of high technology requiring long term development, the organisation of an expo on western technology, and the establishment of a Western Canadian Open University.

Further improvement is also needed in relationships with the Federal Government, and the report recommends further co-operation in: decentralisation of national programmes based on demonstrated performance in fields where western provincial capabilities are strong; tasks of international scope (e.g. opening international markets and scientific relationships); the provision of generically useful information (e.g. competitive assessments of specific markets and technologies, and a rigorous data base for planning in human resources).

Finally a series of three selected policy measures are proposed for each province, to demonstrate what might be done within a system view of Western Canada, taking into consideration the strengths and aspirations of each of them.

The value of Western Canada's human resources and pioneering culture is repeatedly stressed throughout the report. Regional pride, long term vision, innovative spirit and technological sophistication are the difficult prerequisites of a knowledge-intensive economy and Western Canada is impressive on all counts. With further sacrifice (e.g. private funding of educational and research infrastructures), social innovation (e.g. encouragement of women as a source of managerial and technological competence), leadership (e.g. establishment of Innovation Commissioners), the promise of this foundation will be more fully realised.

I. INTRODUCTION

1. Towards a knowledge-intensive economy

Western Canada comprises four provinces: from east to west – Manitoba, Saskatchewan, Alberta and British Columbia (see Figure 1 thereafter). Their total population amounts to some 7.25 million which is 30 per cent of the population of Canada. Together, these four provinces account for 35 per cent of the Canadian gross national product.

Their economy has been largely based and is still very dependent on the exploitation of natural resources. For the whole region, the natural resource-based sectors (including agriculture) employ some 12 per cent of the active population and contribute to 25 per cent of the domestic product. A large proportion of provincial exports is derived from those natural resource goods and materials.

While natural resource-based sectors are still generating significant revenues, their importance is steadily declining in most provinces. Some sectors – e.g. mining, forestry – are facing strong competition from third world countries with lower wages, and the provincial economies appear vulnerable. Moreover they are subject to factors, such as price fluctuations, which are out of control of the provinces. So the western provinces need to enter the new technological era with resolve. This is a basic condition for sustaining long term growth and reducing unemployment – which varies from 6 per cent in Saskatchewan up to 15 per cent in British Columbia.

This new era is characterised by the rapid emergence of a set of new technologies: micro-electronics, new industrial materials, and biotechnologies are the most frequently cited. These technologies are leading to the development of new economic activities (e.g. computer industries and services) and are diffusing throughout the whole range of existing activities in industry, agriculture and services, renewing myriads of products and processes.

These new technologies are leading to a massive intellectualisation of productive activities, requiring huge investment in research and education. Those sectors which are particularly R&D intensive are called "high technology sectors"[1]. Meanwhile the development of the world economy and fierce competition in international trade require enterprises to intensify their efforts in marketing, design, finance, etc. Today, the most progressive firms invest in "software" in the broad sense (research, training, marketing, organisation) as much as in "hardware" (equipment, machinery, building).

This whole process can be characterised as the development of a knowledge-intensive economy. The western provinces must cope with this trend, as must OECD countries. The conversion process is well underway, as shown by several indicators: the growth rate of high technology communities is encouraging; the growth rate of research and development expenditures is higher in the western provinces than it is in the rest of Canada; and new technology-based products are of increasing importance in their exports.

The major goal of innovation policies in Western Canada should be to accelerate this process. Boosting it efficiently requires a broader perspective including further support for the

Figure 1
CANADA

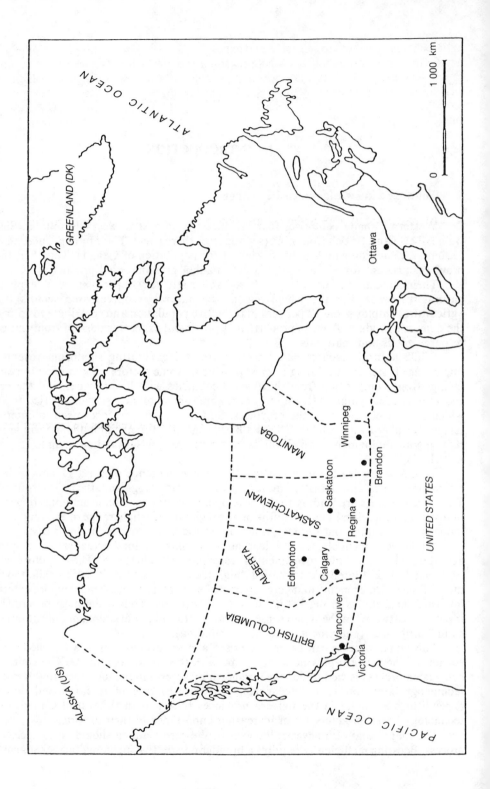

high technology sectors to reinforce and expand them, and stimulation of the diffusion of the new technologies throughout the whole economy and notably in the service sector. Advantage should be taken of the natural resource-based sectors, including agriculture, which should be well "fertilized" by the new technologies. Each province should build its new form of economy on its strengths, avoiding false debates between "high tech" and "low tech". There is just a single notion: "good tech", the one which provides employment, exports, growth, and wealth. It is within this perspective and global strategy that the concept of innovation is taken in this report.

2. Western Canada: unity and diversity

As requested by those who commissioned this study, the western provinces have been examined as a unit. This appears judicious for a number of reasons. The population of Western Canada is, for example, just as large as some regions of the larger European countries, although the smallest western province is larger than the biggest European countries in geographical area. As well, due to commonality of history, problems and objectives, the western provinces need to draw upon the same basic sets of innovation policy measures. Finally, it should be stressed that in a number of fields there is a need for joint inter-provincial initiatives, and united approaches would be most appropriate. So it is meaningful to approach them as a single unit.

However, to a certain extent, it is a fiction to consider Western Canada as a unit. There are important differences between the western provinces as regards their natural resource and manufacturing specialisations. Their conditions of urbanisation and population settlements vary greatly. Their sociological profiles are quite different. It is also true that the future of Western Canada depends mainly on what is done by each provincial government, as joint inter-provincial co-operation is still embryonic. For all these reasons we have felt the need to provide, in the course of the report, specific analysis and policy suggestions related to the situation of each province.

Moreover, the development of the western provinces should be put within the federal or national perspective. The Federal Government has played a substantial role in the development of scientific, technological and industrial capabilities of the West and will continue to do so. The constitutional division of power in Canada is such that both Federal and Provincial Governments have responsibilities for economic development. The Federal Government has exclusive responsibilities for monetary, commercial and competition policies that can influence the investment climate. The Provincial Governments on the other hand have direct responsibilities in such areas as health, education and welfare. However, in many areas the division of powers is not clear cut. There are also policies that are jointly determined by the Federal Government and one, several, or all provinces.

3. Outline of the report

This report, which is addressed primarily to policy makers, attempts to provide critical insights on the climate for innovation in Western Canada and to indicate the types of approaches which would contribute to improve current government policies.

Chapter 2 outlines the economic challenge that the western provinces are facing and provides data on high technology and R&D activities.

Chapter 3 sketches out the main contents of policies developed by the governments.

Chapter 4 considers how the human resources – the basic element of a knowledge-intensive economy – are developed and mobilised in the western provinces.

Chapter 5 reviews support provided to new industry: services for entrepreneurs and innovators, financing available for innovation, and research facilities and programmes. Issues raised by the development of innovation programmes in economies dominated by natural resources industries are also discussed.

Chapter 6 gives some views on the growth strategies to be adopted in relation to new technologies, taking into account existing strengths in each of the provinces.

The main policy recommendations of the study are presented at the end of the report. They include specific policy measures for each province and suggestions of inter-provincial co-operative actions.

An Annex to the report provides a detailed description of the science, technology and industry policies currently implemented in the western provinces and at the federal level.

II. THE WEST IN TRANSITION

1. The economic challenge[2]

There are approximately 7.25 million inhabitants in the western provinces, distributed as indicated in Table 1. Significant increases in population over the last decade (1975-85) have been noticeable in Alberta (34 per cent) and British Columbia (17 per cent) while the growth has been more moderate in Saskatchewan (10 per cent) and in Manitoba (4 per cent).

Historically the exploitation of the natural wealth – wheat, wood, fisheries and more recently oil, gas, coal and minerals – has been the dominant economic activity and the main source of growth. Although the importance of the primary sectors has been steadily declining, the economy of each province is still very dependent on them (Figure 2). The primary sectors provide, directly and indirectly, one out of three jobs in Western Canada.

The importance of agriculture is particularly noticeable in Manitoba (7 per cent) and Saskatchewan (11 per cent). Energy (hydro-power) and mining (nickel, copper) contribute 7 per cent of Manitoba's GDP. Energy and mining (potash, uranium) accounts for 18 per cent of Saskatchewan's GDP. In Alberta, energy and mining (and notably oil and gas production) represent more than 30 per cent of GDP. In British Columbia, forestry, mining, fisheries account for almost 20 per cent of GDP. Most of the western provinces' exports are these raw materials. Their share of the total exports exceed 65 to 70 per cent in Saskatchewan, Alberta and British Columbia. Meanwhile the manufacturing sector has a low weight in the economic structure, representing some 12 per cent of the GDP for the whole region and a small part of the exports (less than 15 per cent). There is, moreover, a strong imbalance in the manufacturing sector in favour of products derived from materials' processing (Table 2).

Developments in the world economy during the last decade have been favourable to the primary sectors. In 1985 GDP per capita in the whole region was now slightly over the national

Table 1. **Gross domestic product at market prices, and population – 1985**

	GDP (C$ billion)	GDP/Cap (C$ thousands)	Population (million)
Manitoba	18.0	16.81	1.07
Saskatchewan	17.3	16.99	1.02
Alberta	62.0	26.28	2.36
British Columbia	54.1	18.76	2.88
Canada	479.4	18.91	25.36

Source: Statistics Canada.

Figure 2

BREAKDOWN OF GDP — 1984

Western Provinces of Canada

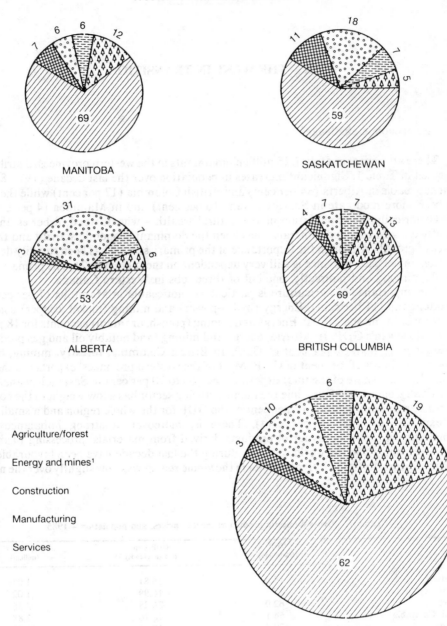

Legend:

- Agriculture/forest
- Energy and mines[1]
- Construction
- Manufacturing
- Services

1.   Including utilities.
Sources: OECD Secretariat and Statistics Canada.

Table 2. **Industrial activity by sector – 1984**

Percent of total manufacturing

	Manitoba	Saskatchewan	Alberta	British Columbia	Canada
Food and beverage	20.4	21.8[b]	18.5	14.5	–
Rubber and plastics products	2.4	0.4	2.0	1.3	3.2
Clothing	7.7	1.0	1.3	1.1	3.1
Textiles	–	0.3	–	0.5	2.5
Wood	2.9	4.5	5.5	24.1	4.5
Furniture	2.4	–	1.1	0.7	1.7
Paper	4.9	–	5.3	19.7	8.4
Printing	9.1	4.8	8.2	5.0	6.1
Primary metals	6.5	–	8.0	6.7	7.9
Metal fabricating	5.7	7.0	7.3	5.5	6.6
Machinery	8.3[a]	9.0	7.0	3.1	4.2
Transportation	12.1	2.0	1.8	4.6	14.0
Electrical products	8.4	7.0	2.4	-2.2	7.6
Non-metallic products	3.0	5.7	5.7	3.2	3.0
Petroleum and coal products	–	–	7.2	3.0	2.4
Chemical products	3.1	5.3	15.5	3.3	7.0
Miscellaneous	1.6	1.2	2.0	1.3	2.8
Value added/GDP at factor costs	12.1	5.0	6.4	13.4	
Total value (C$ million)	1 821	783	3 668	6 107	76 233

a) 1983.
b) Food.
Source: Statistics Canada. Inventories shipments and orders in manufacturing industry.

average (Table 1), and the unemployment rate in each province was less than the national average (12 per cent), with the exception of British Columbia where it has been traditionally higher.

The natural wealth which is still in the soil is phenomenal[3]. But the future development of most of the primary sectors in all the provinces will be constrained by a series of factors. The price of cereals on world markets is expected to decrease with potentially serious fluctuations. The demand for, and price of, hydrocarbon energy are decreasing significantly, at least in the

Table 3. **Exports by destination – 1983**

Percent of total exports

	Saskatchewan	Alberta	British Columbia	Manitoba
USA	34	75.6	47	80.1
Japan	9.5	5.5	25	1.8
UK	2.6	0.5	4	2.3
Other European economic countries	6.1	1.8	8.4	5.3
China	10.7	1.5	2.7	0.2
Australia	0.4	0.3	1.8	0.5
Others of which :	36.76	14.6	11.1	8.3
USSR	15.7	4.4	–	1.0

Source : Provincial Bureaux of Statistics.

Table 4. **Exports by products**

C$ billion

	Alberta (1983)	Saskatchewan (1984)	British Columbia (1984)	Manitoba (1983)
Timber	0.104	0.034	2.667	0.065
Paper pulp	0.207		1.76	–
Coal	0.408		1.53	–
Crude oil	2.25	1.714		–
Potash		0.686		–
Newsprint			0.641	0.46
Natural gas	3.614		0.43–	
Aluminium ingots			0.39–	
Copper ingots			0.38	–
Fish products			0.37	0.001
Grains	2.040	3.172		0.166
Meat products	0.071			0.145
Sub-total	8.256	5.6	8.17	0.323
% of total exports	72	78	70	20
Manufactured goods *of which*:	0.467	0.768		
Chemicals			0.20	–
Machinery			0.19	0.17
Food			0.19	0.06
Communication equipment			0.081	0.02
Petroleum products	0.96			–
Sub-total	1.127	0.798	0.766	0.25
% of total exports	12	11.1	7	16
Total	11.4	7.186	11.659	1.611

Source: Provincial Bureaux of Statistics.

short term. (This adversely affects Alberta which has benefited particularly from the "oil boom", and is now suffering from rising unemployment, increasing outward migration and more generally lower growth.) The world demand for minerals like potash, uranium, nickel, etc., is going to increase steadily but cannot be over-exploited, despite the dominant position held by western Canadian producers. The future of the forestry sector is not very promising (notably in British Columbia, because of poor planning of forest exploitation). More generally, specialisation in the natural resource-based sectors means strong dependence on economic forces which are out of the control of the western provinces, and there is a need for a more diversified economic base.

In fact the transition is well under way. During the last decade, the secondary and tertiary sectors have enjoyed a higher growth in the West than in the rest of Canada. From this point of view present trends point to a reduction of the gap with eastern provinces. Growth of the manufacturing sector in the last two decades has been substantial in the prairie provinces, boosted by growing urbanisation and the extension of towns. During the same period, as in

Table 5. **The micro-electronics-based industry in British Columbia – 1984**

A. ELECTRONIC MANUFACTURERS' ASSOCIATION MEMBERSHIP SURVEY

	Number of companies
Number of staff	
1-10	54
11-50	31
51-100	3
More than 100	6
No data given	27
Total responses	121
Value of sales (C$)	
0-250 000	14
250 001-1 000 000	19
1 000 001-10 000 000	26
More than 10 000 000	14
No data given	48
Total responses	121

B. ESTIMATES FOR THE WHOLE INDUSTRY

	Gross sales (C$ thousands)	Employment	Salaries and benefits (C$ thousand)
Manufacturing	390 000	4 313	115 000
Distribution	58 380	302	9 828
Consulting/services	20 515	218	8 144
Total membership	468 895	4 833	132 972

Source: Economic Benefits of Promoting Micro-electronics through an NSERC II Center – A British Columbia Perspective, August 1985.

other countries, three quarters of new employment has been created in the service sector. The contribution of this sector to total employment (68 per cent in 1980, against 58 in 1961) and to total output is now similar to the average for the rest of Canada, showing that recent structural change in the western part of the country owes less than is usually thought to the dominance of natural resources. Meanwhile natural resources, as a proportion of exports, are decreasing while finished and semi-finished products are gradually increasing in importance.

2. The high technology-based industry

The high technology sectors are the spearhead of the new economy which is now taking form. The development of telecommunications has provided the main platform for the growth of the high technology sectors. The needs for high performance transmissions between remote areas and under difficult conditions has provided a continuous market stimulus to the

Table 6. **High technology industries in Alberta – 1984**

	Number of companies	Annual revenue (C$ million)	Number of employees
Seismic[1]	50	300	3 300
Telecommunications[2]	5	250	1 700
Industrial instrumentation	125	200	2 500
Services	20	120	1 000
Software companies	75	30	500
Total	275	900	9 000

1. The seismic industry figures exclude field staff.
2. The data excludes in-house employees of Alberta Government Telephones and of the major oil companies who are undertaking activities which would otherwise fit into the above classifications.
Source: Data given by Mr. W.D. Croft, Calgary Research and Development Authority, January 1985.

development of a strong telecommunication industry. Another important area of specialisation has been scientific instrumentation related to the needs of natural resource sectors. The growth rate of high technology-based communities is encouraging, although admittedly their relative size is small – compared for instance to the high technology-based community which has grown up in Kanata near Ottawa (25 000 persons).

Statistics on the high technology-based industries are available only for three provinces, and are not comparable:

- The most important high technology-based industry has grown up in British Columbia. The core of the industry is the electronics/telecommunications sector, which in 1984 contained some 250 companies, employing some 5 000 persons, with gross sales amounting to almost C$500 million (see Table 5);
- The high technology-based industry in Alberta comprised 275 companies in 1984, providing 9 000 jobs, with an annual revenue of C$900 million. The seismic and industrial instrumentation sectors, which developed in response to the needs of oil and gas activities, constitute the strongest part of the industry (see Table 6);

Table 7. **Advanced technology companies in Saskatchewan – 1986**

Branches	Number of companies
Agricultural equipment	14
Biotechnology	6
Chemicals	10
Communication equipment	12
Electronics/computer hardware	23
Instrumentation	11
Software companies and computer services	27
Others	7
Total	110

Source: OECD Secretariat, based on information provided in the *Technology Transfer Catalogue*, Saskatchewan Science and Technology Department, 1986.

Table 8. **Gross expenditures on research and development**[1]

1985	GERD (C$ million)	GERD as % GNP	GERD per capita (C$)
Canada	6 530	1.4	257.3
British Columbia	450	0.9	156.3
Alberta	584	0.9	213.3
Saskatchewan	152	0.9	149.0
Manitoba	195	1.1	182.0
United States	135 734	2.8	569.8
Japan	48 900	2.8	404.9
France	17 791	2.3	322.5
United Kingdom	17 583	2.3	309.7
Germany	24 144	2.7	395.6
Australia	2 689	1.1	172.9
Ireland	232	0.8	65.3
Finland	1 048	1.5	213.8

1. GERD = Gross domestic expenditures on R&D. Expenditures in human and social sciences are excluded from GERD for Canada, the western provinces and the United Kingdom.
Source: OECD.

- Saskatchewan high technology-based industry increased from 38 active companies in 1982 to 135 companies in 1985. Sales (estimated) increased from C$81 million to over C$400 million in 1985. Employment in the industry increased from 1 700 in 1982 to 2 900 in 1985. Composition of the industry is given in Table 7.

While the knowledge-intensive sectors are gradually taking the place of the natural resource-based sectors as the dynamo for growth in the West, the global effort in research and development is still modest:

- The western provinces' gross expenditures in R&D, measured as a percentage of GDP, are inferior to Canada's average and other OECD countries (see Table 8). These modest performances are correlated with the economic importance of the natural resource sectors. These sectors, traditionally, invest relatively little in R&D compared to their turnover, and yet they account for a significant amount of the total

Table 9. **Funding of regional R&D**

C$ million

1985	Saskatchewan	Manitoba	Alberta	British Columbia	Canada	Ontario
Provinicial government	13	12	101	27	111	431
Federal government	73	113	150	180	1 128	2 289
Industry	42	21	191	155	1 615	2 721
Universities	15	35	107	52	242	594
Others	9	14	35	36	275	495
Industry as percent of total	27.6	10.8	32.7	34.4	47.9	41.7
Invest. R-D/total invest.[1]	2.4	5.8	3.1	3.1	8.6	5.4

1. This ratio relates the gross expenditures in R&D to the total capital and repair investments (compiled by the OECD Secretariat), year 1984.
Source: Statistics Canada.

Table 10.   **GERD by sector of performance and by province – 1985**

C$ million

Province and region	Federal government	Provincial government[1]	Business entreprise	Higher education	Private non-profit	Total
British Columbia	109	15	195	128	3	450
Alberta	79	46	252	207	0	584
Saskatchewan	45	8	50	49	0	152
Manitoba	88	8	21	73	5	195
Western provinces	321	77	518	457	8	1 381
Central provinces	893	126	2 754	961	81	4 815
Atlantic provinces	161	9	58	104	2	334
Canada[2]	1 375	212	3 330	1 522	91	6 530

1. Includes provincial research organisations.
2. Includes the Yukon and Northwest territories.
Source: Statistics Canada.

R&D effort (35 to 50 per cent depending on the Province). The proportionately larger R&D expenditure in Manitoba is principally due to the federally funded Atomic Energy Whitshell Nuclear Research Establishment, where R&D expenditures are in the range C$70-80 million, representing about 50 per cent of all R&D expenditures in Manitoba;

– The contribution of the business sector, both as R&D funder and performer (see Tables 9 and 10), is weak compared to Canada as a whole and to other OECD countries. This has much to do with the relatively small importance of the manufacturing sector in the whole economy;

– The R&D investment related to total investment appears also weak, compared to Canada as a whole (5.4 per cent).

III. POLICY TRENDS

Before presenting the policies developed both by the Provincial and the Federal Governments to promote technology in the West, it is useful to discuss some general principles which have inspired government action in relation to the culture and society of Western Canada.

1. Socio-cultural background

There are several basic elements in the value system of Western Canada as a region: frontier individualism, perseverance, a forward-looking perspective and distrust of heroes. These values are strongly anchored in the geography of the area and the history this geography has inspired. These values form the common backdrop for the unique ways in which each province has approached its development.

Geographic scale dwarfs the individual at the same time as it inspires confidence in the unlimited status of natural resources. Given the scale of the region, it would have been foolhardy to believe that Government could have any more than a background impact on an individual's life. In short, Western Canadians are used to being in charge of their own welfare and do it well.

The main contribution of Government to Western Canadian history is a significant one: transportation infrastructure. The transcontinental railroad opened the West for development and the network of highways continued to be a major preoccupation of Provincial Government for many years. The railroad opened the area, Government gave people some incentives to take the risk of settling there (venture land, rather than venture capital), but then they were left to their own resources to survive and prosper.

The fact of scale has had an impact upon the industrial structure of Western Canada. The immensity of the territory in which natural resources occur suggests that large industrial companies will be necessary for their development. In short, the economic strength of the large natural resource firms in Western Canada has been welcome. For much of the history of the area, these firms have been the major source of economic welfare beyond that of the individual farmer. It should be noted that a significant number of those firms have been publicly owned: the "Crown Corporations".

In addition to scale, large parts of Western Canada are characterised by another extreme: frigid winters. Not only were the pioneers on their own, but they faced extreme challenges to survival for a good part of the year. Yet survive they did, because they persevered. They did not give up and they did not innovate their way around the problem. They just kept at the job by slow, incremental additions to their property and knowledge.

While nature is awesome in Western Canada, it is also reassuring. Spring does come and the perseverance has paid off. The result is a quiet, but fundamental confidence in the people. They are a forward-looking people, willing to invest in and build towards a better future.

25

In short, the basic premises are that the future is promising if people work hard and steadily as individuals, that no one can be a hero in the face of the immense scale and challenges of the task, and that government can be helpful in providing the common infrastructure needed by individuals.

These cultural premises support the idea of forward-looking investments in innovation and at the same time make it difficult to formulate an innovation policy in which Government plays a strong and direct role: e.g. Government R&D to generate ideas for industry, co-operative R&D programmes between Government and industry in attractive areas, government management of major programmes, direct investments in promising ventures, etc. Instead, the role of Government is most understandable and supportable when it builds infrastructure which supports individual effort: e.g. through research centres, training programmes, venture capital incentives and incubation services for entrepreneurs. Western Canadian Governments have developed their innovation policies on this basis.

2. Provincial policies

Leaders in Western Canada have won public support by arguing that funding infrastructure is investment in the future, that new firms based on technology will generate new jobs (which are more stable than in resource industries) and that stronger universities mean a more promising future for Canada's youth. However, there are also goals which are unique to Western Canada. Two goals which appear across the West relate to regional pride and the stewardship of natural resources. The first is important: westerners want to prove that the West is an intellectual frontier of Canada, and are seeking – already with some successes – to establish new ventures in the attractive fields of micro-electronics, biotechnology and computers. The goal of good stewardship of natural resources relates to regional pride (building on unique strength) and to the need to care for the present while building for the future. These goals of new jobs, a promising future, regional pride, and resource stewardship are the underpinning of the innovation programmes we have reviewed and they structure the directions of public policy.

While the western provinces share many goals in seeking a knowledge-intensive economy there are goals which are unique to each province's social and geographic setting.

The Province of British Columbia is almost as close to Japan as to the Atlantic coast of Canada[4] and its unique geographical position is reflected in the mentality of its inhabitants who, like those on the West Coast of the United States, or in other Pacific countries, support higher education, cultural openness and "laid back" individualism. Provincial Government efforts have been mainly devoted to the support of university research, development of scientific parks and the provision of venture capital for innovation and new businesses.

Alberta is unique in that it is blessed with extensive reserves of petroleum. The wealth generated by this resource has led to an appreciation of technology and an ability to fund it with significantly large amounts of money. The development of innovation policy has been recently stimulated by the publication of an important White Paper on Industrial Strategy for the 1990s[5], which outlines new avenues for the future and helps to determine priorities for further government support to R&D and new technology. Moreover a Department of Technology, Research and Telecommunications has been recently established (February 1986).

Saskatchewan, until recently a decentralised agricultural society, is a more recent supporter of innovation. A Department of Science and Technology was established in 1983. With limited resources in personnel and money, it has focused its effort on the high technology community, providing a comprehensive package of support for entrepreneurs and promoting its image throughout the country and abroad.

Manitoba with a more developed traditional manufacturing base, and especially conscious of its low population, places emphasis on the transfer of technology into existing industry in the province, as opposed to other provinces which emphasize provincially developed technology in high technology fields. A gradual mobilisation of energy around the Department of Industry, Trade and Technology, and involving various groups from university, industry and labour unions has led to the design of a strategy including both the adaptation of new technologies and the promotion of advanced technologies in specialised niches.

The funding of provincial R&D and innovation programmes ranges from some C$10 million per year (Manitoba) up to some C$100 million (Alberta). Agreements recently reached with federal authorities allow significant expansion in the scope and scale of schemes for Saskatchewan and British Columbia (see section 4).

Constraints on the future development of these programmes may appear as a consequence of budget deficits both at provincial and federal level. But problems might be minimised as the financing has been judiciously linked to major funds supported by earnings from the natural resource-based sectors: the Heritage Saving Trust Funds in Saskatchewan and Alberta and other funds such as the Manitoba Jobs Fund.

Although the programmes of each province vary in focus, scope and scale (see Annex), all include some basic components which can be summarised as follows:

– Establishment of technological infrastructures taking the form of specialised centres in the field of new technologies. Some of these have been jointly funded with the Federal Government (see below Federal Technology Centers);

– Provision of support to innovators through establishment of incubator spaces (low cost facilities with common services for entrepreneurs), financial aid for R&D projects, schemes to foster and support co-operative projects between university and industry, etc. It should be mentioned that direct financial support is fairly modest in all the provinces;

– Incentives for venture capital taking the forms of tax credit schemes for investors, or of government capital contributions to specialised institutions. To complement this, favourable tax regimes have been established in each province (see Table 11).

As regards education, Western Canada has a good infrastructure, possessing about one-third of the universities within the country (13 in total). However the educative effort appears lower than in the country as a whole, according to several indicators: level of full time

Table 11. **Major provincial tax rates – 1985**

Per cent

	Corporation tax	Small business	Retail gasoline	Retail sales
British Columbia	16	8	20	7
Alberta[1]	5-11	0-5	0	0
Saskatchewan[2]	14	0-10	0	5
Manitoba	16	10	16	6
Ontario	14-15	0	20	7

1. Alberta has reduced its rate to zero per cent for small businesses undertaking manufacturing and processing, and down to five per cent for other manufacturing businesses effective April 1985 for five years.
2. Saskatchewan has reduced its corporate tax rate to zero per cent on small manufacturing businesses.
Source: Tax Policy Branch, Alberta Treasury.

Table 12. **Education indicators**

		Manitoba	Saskatchewan	Alberta	British Columbia	Ontario	Canada
Expenditure on education	1981	7.7	6.7	5.1	6.2	6.5	7.3
as % of GDP	1982	8	7.2	5.5	7.0	6.9	7.8
Full time post-secondary	1981	16	14.4	14.7	14.6	22.2	20.5
enrolment as % of the	1982	17.4	15.6	15.8	15.5	23.3	21.7
18-24 age group							
% of population aged 20	1981	20.3	14.3	21.7	23.1	23.3	21.9
and over with a degree or							
post-secondary diploma							
University degrees granted	1981	29.2	20.3	14.6	13.9	28.5	22
per 100 persons of the	1982	23.7	20	13.8	14.5	28.5	22.2
20-29 age groups							
Sponsored research in	1982	33	9	24	18	26	23
university per capita							
(C$ thousand)							

Source: Statistics Canada and Ministry of Supply and Services, 1985.

post-secondary enrolment within the 18-24 age group, percentage of population aged 20 and over with a degree or post-secondary diploma (British Columbia excepted), and number of university degrees granted per 100 persons of the 20-29 age group (Manitoba excepted) (see Table 12 data which, however, refer to the early 1980s; moreover, data on the inter-provincial mobility of students and workers within Canada as a whole would be necessary to have a more accurate picture of the education level of the labour force in each province).

Serious financial constraints bear on provincial support to universities across Canada. Since the mid-1970s university funding per student has declined by about 30 per cent in real terms. The western provinces, and Alberta in particular, have made a greater effort than the country as a whole, in absolute terms (grants per students) (see Table 13). It should also be noted that under the so-called EPF (Established Programs Financing) agreements, the

Table 13. **Provincial support to university**

	Level of provincial operating grants per student		Provincial operating grants plus student aid as a % of gross public expenditures		Total university operating expenditures as a % of provincial gross demestic product	
	All Canada = 100					
	1974-75	1983-84	1974-75	1983-84	1974-75	1983-84
British Columbia	105.5	107.8	4.87	3.91	0.92	0.86
Alberta	116.5	127.6	5.42	3.22	0.92	0.84
Saskatchewan	103.6	108.5	4.97	3.90	1.04	1.07
Manitoba	90.4	99.2	5.66	4.16	1.16	1.21
Ontario	93.7	88.8	5.18	4.18	1.02	0.92
Canada	100	100	4.99	3.99	1.08	1.06
(C$ thousand)	2.57	5.63				

Source: OECD Secretariat on the basis of data provided in the "Seventh Report of the Tripartite Committee of Inter-provincial Comparisons (University financing)", April 1986, Ontario Ministry of Colleges and Universities.

Federal Government transfers cash and tax "points" to the provinces to help them pay for health and higher education. These federal funds, however, are not specifically earmarked for universities, and further were reduced as part of anti-inflation initiatives in 1982-84.

3. Federal policies

At the federal level there are three major government bodies concerned: the Department of Regional Industrial Expansion (DRIE), the Ministry of State for Science and Technology (MOSST), and the National Research Council (NRC). There are also several granting councils: the Natural Sciences and Engineering Research Council (NSERC), the Social Sciences and Humanities Research Council (SSHRC) and the Medical Research Council (MRC). Moreover, several sectoral departments (agriculture, energy, environment and others) have significant R&D programmes, either in-house or by contracting out.

As regards science and technology expenditures in each province, the contribution of the Federal Government is significantly more important than the Provincial Governments' contribution, except in Alberta (Tables 9 and 10). The share of federal R&D expenditures (both intramural and extramural) going to the western provinces is approximately 22 per cent of the total expenditures made in Canada, a share which has been stable over recent years (1981-85).

Most western provinces have incurred some funding shortfalls, calculated as the difference between anticipated federal science expenditures based on individual province's contribution to Canada gross domestic product and actual expenditures. This situation can be explained by the fact that the manufacturing base in the western provinces is relatively smaller than that of the rest of the country.

Total federal support for industrial R&D and innovation (including tax measures) amounts to some C$1.2 billion for the country as a whole, equivalent to approximately half of the total private expenditures. Federal support includes:

- A series of grant schemes subsidising R&D projects (see Annex) amounting in total to some C$500 million. Data indicating how they benefit the western provinces are not available, with the exception of the Industrial Research Assistance Program (IRAP). In 1985, 1 214 projects supported in the western provinces received C$12.6 million, representing 21.6 per cent of the total money distributed over Canada;
- Procurement contracts administered by the Department of Supply and Services amounting to some C$300 million in 1985. Western firms have received 10 per cent of this total;
- Tax incentives of various forms. The Canadian Income Tax Act allows the taxpayer to deduct all current and capital expenditures in R&D. In addition, a basic tax credit of 20 per cent is allowed for scientific research expenditures (small businesses are eligible for a deduction of 35 per cent). A Scientific Tax Credit Scheme (SRTC) was introduced in 1985 for investments made in the form of shares, debt and royalty interests for both individual and corporate investors. The effective rate of the credit was 50 per cent, but this incentive was withdrawn at the end of 1985. The value of the total tax support available for R&D was estimated to be worth approximately C$325 million in fiscal year 1984-85.

As regards research and technology infrastructures, federal policies include mainly:

- Support for federal R&D laboratories which are mainly concentrated in the field of agriculture, communications, energy, environment and defence. In the western provinces, federal laboratories employ some five thousand people with a budget amounting to some C$280 million in 1982-83;

- Support given by DRIE to establishment and development of Federal Technology Centers. There are some 30 such centres all over the country, most of them having been created in the last 10 years, jointly funded with Provincial Governments. One-third of those centres is located in the western provinces (see Annex for the details); and

- Support given to university research. Through its three granting agencies (NSERC, MRC and SSHRC), Ottawa distributes about C$500 million annually to support research, scientific training and equipment purchases in about 40 Canadian universities.

4. The Economic Regional Development Agreements (ERDA)

ERDA agreements provide a framework for co-operation and consultation between the two levels of government on matters relating to economic development in each province. These ERDA agreements form a significant part of the Provincial Government strategies, involving large amounts of money – several hundred million dollars for certain provinces over a five-year period, generally funded jointly on a fifty-fifty basis by the province and the Federal Government.

They provide an umbrella for a series of sub-agreements covering a large variety of fields such as mining, agriculture, forestry, tourism, industry, small business, etc. They emphasize activities where the provinces have comparative advantages but where adjustments are needed (e.g. forestry in British Columbia) or areas of possible future specialisation (e.g. transportation in Manitoba). They include various types of policy measures funding, for instance, the development of infrastructures, direct support to industry, aid to exports and eventually R&D programmes (see Annex).

More recently, under such umbrellas, several Provincial Governments have negotiated sub-agreements regarding science and technology matters and industrial innovation on the basis of Memoranda of Understanding (MOUs). These identify mutual interests in a range of science and technology subjects, set priority areas for technology development and exchange of information, facilitating access to state-of-the-art technology and promoting its application.

In 1985, science and technology sub-agreements were in place in two western provinces: Saskatchewan and British Columbia. They cover a five-year period. The sub-agreement with Saskatchewan (the first signed in Canada) amounts to C$33 million for five years and includes policy measures aimed at improving the innovation climate (development analysis studies, industry-university collaboration, etc.) and at providing direct incentives to industry (marketing support, first-use risk reduction, bridging capital assistance, etc.). The sub-agreement with British Columbia amounts to C$20 million. Priority areas are computer science, micro-electronics, applied mathematics and robots, biomedicine and more generally technology transfer based on university research.

As can be seen from this short presentation (sections 3 and 4), the role of the Federal Government in science and technology is far from negligible both as provider of funds and as supporter of infrastructure development. However the federal framework and the provincial diversity and autonomy make it difficult to develop a national science and technology policy. Steps have been taken towards achieving better co-ordination, with the holding of a national conference once a year under the aegis of MOSST with the participation of all provincial ministers for science and technology.

IV. HUMAN RESOURCES

The primary wealth of a knowledge-intensive economy is the human resource. Innovative projects are formed by, and based on, people: entrepreneurs, managers, scientists, engineers, technicians and blue collar workers. We shall consider these occupations and discuss major issues that the western Canadian industries are facing, on the basis of what we have seen in the western provinces. We shall also comment on important social aspects affecting the development and the mobilisation of the human resources.

However we shall begin this Chapter with a few remarks on the development of an entrepreneurial and technological culture among the youth of Western Canada, although the tight schedule of the review team did not allow a concerted study of this important topic. Trends in employment, particularly the diminution of job opportunities in conventional industry, mean that young Canadians need to be familiarised with the new technologies and acquainted with the challenge of self employment leading to small firm development. These trends are common throughout the OECD Member countries.

In this respect, the review team was impressed by the extensive programmes to enhance computer literacy in all four provinces. The quantity of equipment on a per teacher or per child basis is impressive: as far as the review team could judge, there were around ten children per computer at worst and four per computer at best. This "investment in brains" constitutes without doubt an investment of huge future potential.

Education systems in the western provinces also have well planned programmes for inserting youth into the real technical world and relating classroom material to practical events. Saskatchewan showed a particularly innovative approach, with a campaign launched throughout the province to give a thousand secondary school students the opportunity to relate to a space mission. The particular mission of a US space shuttle had on board a Canadian crew member, and using remote telemetry systems, the school students were able to make moisture development observations. The impact of this involvement was great, and led to raising of interest far beyond the schools themselves and the families directly concerned. The review team was also impressed by the participation of western Canadian youth in the Canada-wide Young Enterprise scheme, and the support given to such a scheme by western companies.

1. Entrepreneurial resources

In general terms the entrepreneurial culture of Western Canada is good. Western people continue to manifest a strong pioneer spirit. They are risk takers. Incentives for small business – tax incentives notably – are stronger than in most OECD countries. New technology-based firms are encouraged. Entrepreneurs are committed to development of Western Canada, and to their provinces.

The review team met several dozen entrepreneurs/innovators in each province. Those entrepreneurs were generally ex-employees who, having had an idea not necessarily related to their previous employment, had decided to go it alone. There were also several relatively well-established businessmen who were using their finances, built up in the natural resource boom years, in new ventures and products. Several entrepreneurs were former academics who had marketed an idea which had emerged from their research or teaching. Finally, a few entrepreneurial farmers were interviewed.

Major sources for entrepreneurs in all provinces are employees in existing companies, the academic world, and the rural community. These sources have been tapped to differing extents, but are all underutilised. This situation is partly related to an overemphasis on the electronics industry, too much identification of high technology with computers and insufficient awareness of innovation and entrepreneurship (in certain groups, e.g. farmers). In some provinces the social and financial status of industrialists (entrepreneurs, managers, engineers) whose salaries are in average 20 per cent lower than earnings of other groups such as doctors or lawyers may also constitute an obstacle to the development of entrepreneurs.

Given the uncomfortably high levels of unemployment, the unemployed are another pool of entrepreneurial talent. This paradox arises because the unemployed generally receive aid (and advice) that directs them to existing industry or maintains them unemployed instead of encouraging self-employment. Several OECD countries – e.g. the United Kingdom, Ireland, Finland – have successfully introduced schemes providing entrepreneurial incentives and training as part of unemployment systems.

The educational level of entrepreneurs was generally high and many persons met by the review team were qualified scientists and engineers. However, a significant number were insufficiently trained in the financial, managerial and marketing aspects, and thus were experiencing difficulties after starting up their business.

Incubator spaces are an efficient tool for increasing the number and quality of entrepreneurs. These spaces reduce considerably the mortality rate of new companies, notably new technology-based firms (from an usual rate of 90 per cent to 50 per cent). The concept of incubators is well understood in Western Canada, and some of those visited were obviously effective: e.g. Discovery Park in Vancouver (British Columbia) and the Technology Commercialisation Center in Winnipeg (Manitoba). There are, in fact, various types of incubators. Some may be created and run by the public sector, while others are mainly based on private initiative, funding, and responsibility. Some incubators can be focused on high technology entrepreneurs, while others are fitted for more conventional entrepreneurs servicing local needs (e.g. beehive units). Choices between those various options depend on local interests and opportunities.

There is great scope to increase the number of incubators in Western Canada, and each province should establish multiple incubators. One may estimate that a comprehensive incubator programme across the West could elicit, at minimum, 3 000 new technology-based firms[6]. With a mortality rate of 50 per cent, an average of 10 employees per firm and a multiplier effect of 1.6 (one job in high technology generates another 0.6 job, in ancillary activities, e.g. transportation of shipments, printing of pamphlets, etc.), one may estimate that 20 000 to 25 000 jobs can be created across the western provinces over a five-year period. This represents almost 1 per cent of the active population and 10 per cent of the unemployment.

2. Scientific and technical manpower

The workforce of a knowledge-intensive industry includes up to 50 per cent of intellectually trained people, distributed approximately equally between managers in the broad sense (i.e. engineers, scientists, salesmen, etc.) and technicians. The remaining 50 per cent of the workforce includes clericals, skilled workers, semi-skilled and unskilled workers. Due to the lack of sound and comparable data on present and planned outputs of diplomas in the western provinces' education systems, it has not been possible for the review team to assess the extent to which these outputs will meet the future needs of the new technology-based industries, with reasonable growth expectations, e.g. a 10 per cent annual growth rate, corresponding to a doubling in 7 years.

As regards more particularly outputs of engineers, and on the basis of past trend of enrolments in engineering in the most important technical universities (Table 14), it appears that the growth rates are almost satisfactory in Alberta and British Columbia (8 per cent per year) but insufficient in Manitoba and Saskatchewan (3-5 per cent per year). It appears also that the share of engineering students within the total university enrolments has been slightly

Table 14. **Engineering education in Western Canada**

UNIVERSITY OF MANITOBA[1]
Full-time graduate and undergraduate enrolments (winter sessions)

	1979	1980	1981	1982	1983
Engineering (a)	1 231	1 227	1 303	1 398	1 474
Total enrolments (b)	12 157	12 327	13 283	14 646	15 648
a/b in percent	10.1	9.9	9.8	9.5	9.4

UNIVERSITY OF SASKATCHEWAN[2]
Full-time degree enrolments

	1981-82	1981-82	1982-83	1983-84	1984-85
Engineering (a)	1 156	1 178	1 218	1 256	1 297
Total (b)	10 156	10 814	11 845	12 744	13 174
a/b in percent	11.4	10.9	10.3	9.25	9.8

ALL ALBERTAN UNIVERSITIES
Degrees awarded (bachelors, masters, doctorates)[3]

	1971-72	1975-76	1977-78	1979-80	1981-82	1983-84
Engineering (a)	486	371	481	558	622	703
Computer science (b)	86	76	84	141	148	169
Total (c)	7 638	9 997	7 782	7 700	7 201	8 011
a/c in percent	6.3	5.3	6.1	7.2	8.6	8.7
b/c in percent	1.5	1	1	1.8	2	2.1

ALL BRITISH COLUMBIA UNIVERSITIES
Full-time graduate and undergraduate enrolments in engineering and applied sciences

	1980-81	1983-84
Engineering and applied science (a)	2 418	3 106
Total (b)	29 630	37 214
a/b in percent	8.16	8.34

1. Other Manitoba universities do not include an engineering college.
2. Engineering enrolments in the University of Regina are approximately one-third of those in the University of Saskatchewan. Their share as a percentage of total enrolments has not changed over the last five years (5.2 per cent).
3. Enrolment data not available.
Source: OECD Secretariat, on the basis of data provided by the Universities' Yearbooks.

declining in Manitoba and Saskatchewan, and slightly increasing in Alberta and in British Columbia. Strong efforts are needed to "market" engineering and applied science studies among young people, if the western provinces really want to base their future on knowledge-intensive industries, high technology exports and so on.

More generally there is an insufficient awareness among policy makers of the crucial importance of adjusting the education system to future industrial needs or targets. There are very few solid surveys on these critical matters. Two studies were brought to the attention of the review team: one on the electronics industry in British Columbia and another on the technology community of Saskatoon[7]. The surveys point out the gaps that exist between the "demand" and "supply" of qualified people. A large majority of firms visited in the course of the review were beginning to face a shortage of qualified engineers and graduate technicians. This gap is going to increase substantially in the coming years, even when one considers the possibility of attracting good specialists from the rest of Canada or from foreign countries, notably the United States.

It is surprising that students in some provinces should have to go to neighbouring States of the United States to enrol in courses they need for jobs in the western provinces. Some education programmes, directly related to present development of industries which are contributing to provincial strengths, have not been updated. For example there is a near absence of courses in micro-wave technology which is of critical importance to the telecommunication industry. As regards provision of scientists for advanced fields of importance for the western industry, such as biotechnology, serious shortages came to the notice of the review team. An example is toxicology where the number of post-graduate specialists (PhD) trained in Western Canada (and Canada as a whole) does not exceed twenty. This situation appears to be mainly due to absence of planning at both a regional and national level.

In general terms, the importance to be attached to the development of sound higher education bases cannot be too strongly emphasized. The budgetary reductions that a number of universities in the western provinces have experienced is having an adverse impact on this development. Increases in tuition fees paid by students do not constitute a solution to this problem.

The technological institutes visited by the review team were well staffed and well equipped. Finer tuning of their equipment appears, however, necessary here and there, either because it is too sophisticated in relation to needs or because it is outdated as regards most recent developments. This is particularly true in the field of computers and CAD/CAM systems.

The community colleges are training large numbers of technicians and manpower in new technology. They are close to their communities and their industries and can respond rapidly to their needs. These colleges are a real strength and asset to the development of the innovation climate in Western Canada.

Facilities for retraining and recurrent education (mostly based on these institutes and community colleges) are also well developed. The most impressive achievements are in Saskatchewan, where plans have been made to increase training places by 60 per cent in three years (8 600 positions for 1988). A number of apparently well-designed programmes are set up, in close collaboration with industry which participate in their design, thus ensuring that proper emphasis is put on "doing" rather than on "knowing". The only question that must be asked regarding the Saskatchewan technical (re)training system is whether the enterprises should bear more of the cost for the services they obtain. Low charges may present a serious drawback in that firms do not appreciate sufficiently the value of training investment and do not accurately assess the services offered by the training institutes.

3. Managerial competence

An inhibiting factor in the expansion of new high technology businesses seems to be a lack of good managers. This is becoming particularly evident as firms expand to the stage where the original owners cannot between themselves fulfil all the management functions. Finance and marketing in general, and export marketing in particular, are areas of incipient or actual weakness noted by the review team.

More progressive business courses are needed in the community colleges, the technological institutes, and the universities. There should also be systematic attempts to establish programmes combining management and engineering education. Innovative approaches which have been developed by certain establishments in the western provinces should be studied and adopted. For example the University of Regina introduced sandwich courses fifteen years ago which provide formal teaching combined with practical experience. More recently this university has also set up inventive training programmes for engineers which integrate regional and ecological planning.

One of the most important factors in the development of talented managers is learning from mistakes which have driven the firms to and over the brink of failure. In this regard, the propensity of western governments to "bail out" firms is undoubtedly a weakness in their effort to build corps of experienced managers. Moreover each province must have an active programme to locate and track the success of "native sons and daughters" and recruit them back to the province once they have accumulated international experience in technology, marketing, finance and management.

The development of an international perspective should reduce the problems created by the isolation (and inward orientation) of western Canadian communities. This isolation leads to lack of awareness of international competition among entrepreneurs, lack of knowledge of technical progress in school equipment, etc. Various policy measures can be considered to overcome the problem. Some may be of a short term nature: e.g. those schemes offering systematic opportunities to entrepreneurs and innovators to visit abroad and participate on international technology fairs. Some may have a longer term perspective: e.g. exchanges of students and professors with foreign countries on a significant scale. Possible approaches to this include the creation of International Management Institutes like the one contemplated in Alberta, with promotion of long-term student exchanges notably with the Pacific Rim countries.

This inward orientation operates not only at the level of Western Canada in relation to the rest of the world, but within Western Canada itself. To take just one example, innovators in one province frequently appear unaware of technological developments or competitors of direct interest to them, existing in the neighbouring provinces. Establishment of regular competitions or exhibitions in various fields (agricultural technology, computers, etc.) would improve the situation.

Lack of communication created by isolation can be observed among communities in the same province. One province has a long tradition of social, political and educative innovations. In the latter field, for instance, the list of innovative approaches introduced in the school system over the last decades is impressive. The point is that those schemes introduced here and there by specific establishments do not diffuse throughout the province. The diffusion patterns follow the highway routes but do not go further. Strong actions are needed to promote exchanges of experiences, develop awareness and expose promoters of these initiatives to potential adopters not only at the level of the province, but at least of Western Canada. Incidentally this isolation sometimes leads some groups to "reinvent the wheel". This is for instance the case with projects using new technologies such as nodal computerised networks between school establishments. Better performing and cheaper systems operating already in foreign countries could have been imported, probably with a better cost/benefit ratio.

4. Social issues

To conclude this Chapter, three issues will be briefly discussed: industrial relations between the social partners, the status of women in the economy and some ethnic aspects.

a) Industrial relations

Although impacting to a various degree on the different provinces, the relationships between the social partners – industrialists and managers on one side, the labour force and trade unions on the other side – are uncertain. This creates a climate of ignorance or antagonism which constitutes an obstacle to progress and to the introduction of new technology. The climate can be improved only through measures which commit the partners to specific actions in relation to entrepreneurship and innovation. For instance in British Columbia – the province which has been suffering probably the most from these antagonistic relationships – there are several forward looking labour unions who are prepared to fund innovation and high technology development through use of savings in credit unions and pension funds. Further, in the forest industry the labour unions have supported the initiative of awarding small forest quotas to self-employed woodmen who are the source of innovative processes and equipment for the timber industry.

b) Status of women

Male dominance is a striking feature of western Canadian societies. This is perfectly understandable. Conquest of new territories and exploitation of natural resources were mainly men's activities. This situation, however, should change if the new economy is to take shape. Studies made in OECD countries show the increasing contribution that women are making to creation of new enterprises, and show that when women do go into business they are often better at it than men.

Western Canada has greater needs for professional talent than its male population can satisfy, yet it lags behind most OECD nations in recognising the importance of half the available human resources, and in policies and programmes to encourage professional and managerial careers for women. In particular, the review team was impressed by the leadership potential of the women it did meet and by their progressive, yet realistic, perspective on their own culture. For example, British Columbia had a Deputy Minister for Women's Programs for the province who clearly understood such realities as the need for economic data to validate opportunities lost and the need to overcome the backlash from women who had devoted their lives to being "frontier wives"[8]. In short, Western Canada already has the talent needed to correct this situation. What is now required is the same action that we shall stress throughout this report: development of visible leaderships and linking these into effective networks for the formulation and execution of policy.

c) Ethnical fragmentation

A mixed population should on general principles be more vigorous than a monoculture. The western provinces are fortunate in having a great ethnic mix, and unfortunate that the elements in the mix tend to isolate themselves. No group has a monopoly on innovativeness or

36

is necessarily best at it, and the western provinces have a great opportunity to be innovative in ways that other regions or countries have not. Everyday examples can be found showing the benefits of such an integrated approach which utilises the creativity of the population as a whole. A case in point is a discovery made by a team of Eskimos in the North, led by a German scientist who recently immigrated to Canada, that cold water attacks iron more severely than hot water does, a discovery contradicting theories established over centuries. In a knowledge-intensive economy, even the most traditional fields can be improved, and all the creative resources and experiences should be tapped.

Apart from Saskatchewan, the western provinces have not developed large and systematic programmes towards Indian people. In the University of Regina an associated Indian Technical College has been formed and has trained more than 600 persons in ten years. In the training system a comprehensive scheme is implemented including detection, support, financial aid, integration steps for hundreds of Inuits. Such initiatives should be spread and adopted elsewhere.

V. SUPPORT FOR NEW INDUSTRY

Technological innovation and economic growth take place in those regions characterised by:

- The existence of sound and extensive networks and services to support the birth and growth of innovative projects in various aspects: technical, commercial, financial and managerial;
- The availability of sufficient money to finance these projects, particularly in the form of venture capital;
- The establishment of strong research facilities and dynamic programmes which nurture the knowledge basis and attract technologically advanced people and firms.

We shall comment on these various aspects on the basis of what we have seen in the western provinces, pointing out interesting and efficient initiatives on the one hand, and weaknesses on the other hand. We shall also indicate, in the course of the discussion, policy mechanisms of other countries which could inspire policy makers in the western provinces. We shall conclude this Chapter by discussing an important issue: the support provided to the development of the new industry by the established (natural resource-based) industry.

1. Services for innovators

The elements for networks supportive of innovation are at an early stage of development in Western Canada. Federal and provincial networks exist side-by-side in a not always easy relationship. The sources of support are fragmented and, in general, have a low public profile. Awareness of the existence, purpose and value of these networks needs to be raised in the entrepreneurial community.

At present all four provinces have set up organisations to promote and support innovation and entrepreneurship. The present staff are of impressive quality and dedicated outlook. However most of these organisations seem understaffed considering the current number of high technology firms, and are certainly understaffed to take on the task of servicing the increase in clients that would arise from stimulation of demand. On the other hand the physical plant (buildings), infrastructure (library and information sources) and funding levels range from good to excellent. The support organisations appear to be beginning to collaborate among themselves as illustrated for example by the Small Business Association of Canada, but real co-operation and networking is yet to occur throughout the provinces.

Three aspects which should retain particular attention from policy makers will be discussed below:

- Expertise for evaluation and packaging of projects;
- Support for technology and market research; and
- Channels for trading high technology abroad.

38

a) *Evaluation and packaging of projects*

The necessity to evaluate projects, in order to assess their economic and technological viability, is clear to both proponents as well as to backers. The capacity to evaluate innovative projects is to some extent as much an art developed by experience as it is a science founded on well-known rules. An increasing number of regions in OECD Member countries are establishing organisations able to carry out such evaluations. None of the western provinces has such a centre, and at present such evaluations are often sent to the east of Canada – especially the University of Waterloo – as well as to the United States – especially the University of Oregon. The need for such a centre has been recognised in Alberta, and plans are well advanced to establish a publicly funded organisation. A private sector company is being set up in Calgary which includes project evaluation in its services. In British Columbia, networks operating around Discovery Park (see below) provide, to a certain extent, such a function. In other places, too many projects are evaluated only when individual entrepreneurs are contemplating investment.

Because innovative and high technology projects are unfamiliar to the public and to financiers alike, they must be presented in an attractive and substantial way in order to gain support. In particular the quantitative aspects of the project need to be well documented by data founded on technical and market information. Extra support is required by intending entrepreneurs to help them present projects and to put them in touch with sources of skills complementary to their own. Equally there is a need for a mechanism to put entrepreneurs and promoters in touch with sources of viable ideas. Both these functions can be exercised through an innovation centre but this may be oversized especially when smaller projects are concerned.

At the most basic level, intending entrepreneurs need an information package of the type being used in Saskatchewan. Such a package of co-ordinated documents is more useful than occasional leaflets. Behind the package there must be an advice point that can direct entrepreneurs to relevant sources, whether public sector or private sector, and to complement by face to face counselling, the information given in the brochures. Further it is crucial to have a mechanism to help the technical innovator write a business plan in a persuasive style. This last feature cannot be too strongly emphasized.

In many OECD Member countries, there are various forms of training to help entrepreneurs with business plan preparation. This training may be as little as one day, several evenings, or a week-end. It may be provided by the public sector (in continuing education facilities) or by Chambers of Commerce and enterprise agencies. This type of training could be certainly more broadly diffused in Western Canada. Universities could also take the initiative of organising appropriate meetings for the business community such as the "MIT Enterprise Forum" which provides evening or week-end sessions where entrepreneurs find contacts and advices.

Incubator spaces surrounded by appropriate networks constitute also an efficient approach to project evaluation and packaging. An example is provided by the Discovery Enterprise Program of British Columbia. The scheme is a pre-venture capital fund which encourages partnership and joint venture with other interested investors. It encourages them to examine opportunities offered by innovative projects in early stage of development as well as meeting with their entrepreneurs/promoters seeking for finance. Sectoral groups have been established in electronics, manufacturing, biotechnology and agricultural food, etc. These groups meet once a month and examine a dozen projects during the day. This approach is informal and makes the best use of networks. It provides catalysis for bringing together competences and expertises in evaluating projects. The basic principles and methods of this scheme could certainly be adopted elsewhere.

However, for a more systematic approach it is recommended that each province establish evaluation centres, coupled with incubator spaces, diversified by geographic location and eventually by subject focus in relation to technological specialities (see Chapter 6). Based on the US experience, the keys to evaluation centres are that they:

- Are typically based at a university which has the status to pool needed technical and managerial talents and the types of expertise needed by entrepreneurs;
- Include successful entrepreneurs who have "been there" and can give a realistic assessment of the project as a "business" and who can be initial contacts into the networks of suppliers;
- Include financiers, lawyers, accountants and other professional ancillaries who can provide crucial advice on "how to" (not whether to) carry out the project legally and satisfactorily.

b) *Consulting and research on technology and market*

Western Canada has all the access to sophisticated technology and the intelligent professionals it needs to start many businesses in limited market niches. However, a technology which has potential for sales on a continental or world scale is one which must operate reliably and inexpensively in a wide diversity of settings. Brilliant engineers and scientists may formulate the ideas behind such technologies, but it takes the hard work of technologists, consultants and vendors working with internal specialists to turn the ideas into functioning products.

Examples encountered in our field survey include: an innovative tape drive for computers requiring vibration analysis, advanced motors geared to its specifications, and interconnections matching forecasted interface specifications for leading devices; a new biomedical procedure requiring new instruments designed around revolutionary drug-delivery specifications, new assay procedures and micro-instruments and new methods of chemical processing to deliver necessary purity in large batch production. Such developments need a wide range of support services such as prototype design and construction, development engineering and design, materials and component testing and quality control. At present a large number of the western high technology firms are dependent on central Canada and the United States for the provision of such goods and services, and this fact reduces their competitiveness. However, some infrastructure has begun to be developed in the West, e.g. in Manitoba for manufacturing technology (Manitoba Research Council's Center), in Alberta for electronics (with a recently established "Test Center"); moreover, a certain number of private engineering and consulting firms provide specialised expertise. Careful examination is needed to assess the extent to which these infrastructures are responding to the needs of technological entrepreneurs, and those services which are lacking should be established or promoted actively.

There is another reason why firms flock to Silicon Valley or Route 128 or to Kanata near Ottawa. In these locations there is an informal network of innovation consultants and vendors which provides the ancillary products and knowledge needed by the area's entrepreneurs. These partnerships assist the firm not only in reducing its technology to practical form, but also in adapting it to the rigorous demands of broadening opportunities. Management of consuming and producing firms must master the continued learning of such essentials as:

- Which of several possible applications are most significant for the early proving of a technology and how will these grow into mass markets (e.g. a firm which developed an electronic warning system for police and fire services found its first market in the new field of express delivery in the USA);

- Which convergences of existing technologies have synergistic potential for multiplying the benefits from the firm's technology (e.g. converging advances in pharmaceutical, micro-instrumentation and animal genetics are bringing about a breakthrough in a biomedical firm).

These examples show the critical importance of the ability of technology companies to interact continually and informally with vendors of innovations, producers of potentially converging technologies and expert consultants. To the extent that it is true, as argued elsewhere in this report (see below Section 4), that western firms possess a "resource mentality" that encourages them to rely upon technologies already proven by others, then to that extent western ventures will lag in competition with others who possess this interaction. To the extent that the western provinces establish successful programmes to incubate new businesses, but their established and Crown Corporations are reluctant to engage in early use and jointly managed learning, then to that extent will Western Canada give birth to entrepreneurs who, as a result, will be frustrated as others advance beyond their early lead and who are likely to leave Western Canada for more progressive markets. We have met such firms in the course of our visit.

Networking, mutual support and cross fertilization can be reinforced by the establishment of associations (or foundations) bringing together local industrialists and eventually people from other origins (e.g. university researchers). These associations can be created at the level of a whole high technology community, of particular industries (e.g. electronics) or eventually of more specific technological fields. The public sector can stimulate the development of these groups, in providing some financial support.

c) *Commercial networks*

For high-technology products the importance of marketing is quite critical, not least from the comparatively isolated locations of Western Canada. By tradition, however, the natural resources of the provinces have been marketed outside Canada. The mechanisms for this marketing include the producer boards, the wheat pools, the Potash Corp., and similar bodies; the representative trade offices overseas of both Canada and the individual provinces; and the agents and companies in foreign countries that import Canadian natural resources. There is an existing network to service the natural resource industry. There is in principle no reason why this network should not be used to benefit high technology. There are two major facets to be considered. In the first place high technology products related to natural resources can be marketed along the same or related channels. In the second place the natural resources may be traded off against technology (and associated capital).

From the viewpoint of the western provinces an important growth area for exports is the Pacific Rim and statistics highlight this growth for particular countries. Several of these countries are also markets for innovative technology related products, and are also sources of high technology – Korea, Taiwan and Hong Kong as well as Japan. The trade representatives for the western provinces in these countries could, and should, be enjoined to scout for suitable sources of technology that can complement the development strategy in the provinces. When suitable technologies are located, the possibility of exchange deals arises. In such a way the technology content of exports from Western Canada can be increased and markets created.

2. Finance for innovation

a) Government aid

Financial aids to innovation are as varied and fragmented in Canada as in most other OECD Member countries, and range from assistance for research through to risk capital. In Canada, each province has its own system and the federal programmes are executed in parallel. Further, some of the ERDA agreements currently being negotiated between the provinces and the Federal Government include specific new programmes of financial support for innovation and new technology. In many countries where the entrepreneur's viewpoint has been recognised, the trend is to a "single window" where one single organisation handles all the individual assistance programmes for objectives such as technology development, marketing, regional development and export promotion. The "One stop shop" established by Saskatchewan's Department of Science and Technology to deliver its programmes and services and to refer clients to other agencies is in line with this general trend.

However, the business and entrepreneurial community understands neither the plethora of special grants and support mechanisms nor the fragmentation into artifical divisions and sub-programmes when the support is, in reality, for developing the enterprise or a product/process. The review team found that the support that was almost universally welcome in Western Canada was the IRAP programme of the National Research Council. Positive features often cited were rapid decisions on applications and minimal bureaucracy.

Two aspects about the administration of financial aid programmes need comments: the expected failure rate and the span of the innovation process covered. Regarding the failure rate, provincial officials are proud of rates of survival of 90 per cent or more. One-half or one-third of this rate would be more in line other OECD countries, as innovation implies risk and innovators learn through failures. The need to support the whole innovation process from idea generation to production stage in companies is insufficiently understood. Moreover, one may ask whether there is sufficient financial support for development work. Current schemes do not provide significant funding for this phase of the innovation process.

Procurement policies could contribute to reducing the gap in development funds. When totalling the spending from all different sources – government contracts, Crown Corporations and so on – it is obvious that government procurement would exert a powerful leverage effect on the innovation effort in each province. On average, government purchases represent some 15 per cent of the gross domestic product of the provinces. The western governments do not seem to have a systematic and sustained approach to this question, although efforts made in Saskatchewan to publicise on a regular basis contracts for major government projects should be studied and copied elsewhere in Western Canada.

As regards the Federal Government, the role of Government contracts (Department of Supply and Services) in the starting up of a number of new technology-based companies is well recognised. However, a significant number of companies have found difficulties in tendering. There are problems in administration and in the high costs for tendering. The costs for documentation and for long distance travel for presentation tend to discourage candidates, even the most resolute ones. Moreover, sometimes firms far from Ottawa only become aware of what projects are up for bid, after the decision on bidding has been made.

In Scandinavian countries, where large government defence programmes are fewer and where regions are still in stage of development, Governments use a combination of risk loans and grants to support technically advanced projects. The government role in the first stage of development work is more grant oriented, whereas in the latter stages it is more risk sharing: loan guarantees and risk loan finance are appropriate. Approved projects can be financed for up to 75 per cent of their total cost by these methods. Further, if projects are not profitable the "risk loan" element can be converted into a grant.

One of the basic problems in financing technology-intensive projects is that all the projects do not have the same degree of innovation. Some are highly innovative and some have little innovation content although they are relevant for regional economies. But both need financing until the incoming cash-flow from sales can cover the costs. The risks involved in the development of new ventures must be handled as a complete system. The major part of the risk is carried by the equity holders, the entrepreneurs and the venture capitalists, but the Government may have a vital role to play in filling the gap between reasonable equity and the usual risks carried by the banks, insurance companies, and similar conventional lenders.

b) Venture capital

Venture capital is the usual name for equity investment. The amount of investment put by venture capital organisations into industrial development is in fact modest: around 2 per cent of the total. Even in the United States venture capital companies invest "only" in some 1 500 companies annually, while there are some 600 000 company start-ups annually. But, in providing money to the most aggressive and progressive firms in new technology, venture finance plays a key role in the most dynamic part of the economy.

In practice different involvement is possible for venture capital companies: no involvement ("hands off" management), venture nursing, joint venture with involvement in day-to-day management ("hands on"). In Canada all these types are needed and venture nursing is particularly advisable in new technology companies when the entrepreneur must be continually assisted.

The venture capital market in Canada is rather small both in terms of volume of business and number of companies involved. The centre of activity is in Eastern Canada. The membership of the ACVCC (Association of Canadian Venture Capital Companies) is approximately fifty companies, only eight of which are located in the West – one in Manitoba, one in British Columbia, one in Saskatchewan, and five in Alberta. However, there are sources of venture capital in the West that are less formalised than the eastern Canadian corporations. Several examples were noted by the review team. These included family groups in Saskatoon, a group of doctors in Saskatoon, and particularly rich individuals throughout the region, one of whom was notable for having set up his own bank. Furthermore, there are provincial programmes that are so close to providing venture capital that they should be included in the venture capital market. Up to now, provision of venture capital from these various sources has apparently been sufficient to support the growth of high technology.

Moreover all provinces have recently promoted the setting up of Small Business Equity Development Corporations (SBEDC) (or Small Business Venture Capital Corporations) by various means such as tax credit or cash grants for investors in these corporations (individuals and other entities). The SBEDC mechanism has a successful track record in the US, and has been adapted in other countries including Australia and the United Kingdom. The precise detail of financial advantage is unimportant: the main feature is that the SBEDCs allow the aggregation of small packets of investment money into meaningful amounts.

However, as in most OECD Member countries, it is difficult in Canada to raise relatively small amounts of capital. There is, in fact, no formal mechanisms to raise money for small companies defining, for instance, acceptable levels of liability, "fair disclosure" practices, etc., in financial deals. Entrepreneurs seeking for finance are involved in costly and long procedures through provincial securities commissions, which often base their decisions on value judgments, notably as regards "fair deals" between the entrepreneur, the investors and the deal promoter. Thus the availability of small parcels of venture capital is an identifiable gap in the capital markets. In general these smaller parcels are being painfully raised by intending entrepreneurs from their own resources, and by mortgaging their properties, from relatives, and from conventional finance sources (banks, insurance companies, finance houses). This squeeze on the smaller entrepreneur is causing undercapitalised companies to be set up, with some companies carrying too much debt burden. These factors do not promote the rapid growth of technology companies that everyone wants.

The availability of venture capital in Western Canada is doubtless higher than it appears and there are, or should be, plentiful funds that could be invested in innovative high technology businesses[9]. The raising of public awareness and the adaptation of financial practices will increase both the number of entrepreneurs in the market place and the number of investors. To bring these players together needs highly experienced people, whether they operate within formal venture capital companies or through more informal networks. In the context of Western Canada, high technology projects need champions. These champions have to operate in a financial environment that is usually dealing in multi-million dollar projects. They need to be persons of substance with good contacts in the political as well as financial community – in colloquial terms they are "shakers and movers". They are in practice real entrepreneurs. There are such people in each of the provinces, but their number needs to be increased to constitute a critical mass.

A last point to be considered is the relative weakness of the stock exchanges in Western Canada and the comparatively high level of regulation of the securities industry. Without "exit" mechanisms venture investors will be reluctant to come forward. Changes are in hand, particularly in Alberta, and with the comparative strength of the venture capital industry in Alberta it would seem to be a realistic option for a "junior" trading board in that province, to be devoted to, and to serve as a resource for all the western provinces. The long standing trading in thousands of minor resource stocks in Vancouver is probably not a favourable precedent *a priori*, but no doubt arrangements could be made for trading in "junior" stocks of high technology issues in Vancouver. The importance of stock exchanges is greater in Canada because the other "exit" mechanism – takeover of the new technology company by a larger company – is severely limited by the low number of "predator companies" in the country.

3. Research programmes

This section will begin with a discussion of the nature of university and industry linkages. It then reviews the climate for, and status of, R&D in the two fields of micro-electronics and biotechnology, which have received consistent interest, and support, of technology policies in the provinces. We shall then examine the status of R&D in the natural resource sector which is the other main area in which western laboratories are active.

a)   Industry-university linkages

The most important infrastructure element in each province is its universities. Universities are in an important position, therefore, to assume leadership through the production of new technology (spin-offs), leadership in stimulating networks centred on knowledge-intensive industries and leadership in preparing youth for careers in such entrepreneurial fields. Such contributions are new in comparison to the traditional contributions of universities to objective science and basic areas of education. How Western Canada universities meet this challenge of institutional innovation will have significant influence on the region's success in knowledge-intensive industry.

An important commonality across western universities is their awareness of the challenge and opportunity they face and each, in its own way, is attempting to meet that challenge. As might be expected, the newer universities seem to be rising to the challenge with the most energy and innovation (especially in educational innovation and entrepreneurial management), while the older, provincial universities (with the exception of the University of Manitoba) seem to rely on incremental modifications. Most universities have (recently) appointed officers in charge of industrial liaison, or opened offices of industrial research, with the support of Provincial Governments. Nowhere, however, did we find a programme which was comparable in sophistication and capability to those found in US or European technical universities – universities which are aggressively expanding their linkages to industry.

None of the universities had worked out modern policies on the ownership of technology: e.g. rules for the sharing of royalties and equity between faculty and the university, the ownership of computer software, priority rules on publication versus intellectual property protection, responsibilities for infringement prosecution and the use of third-party licence brokers.

None of the universities visited had considered the need for brokers and extension capabilities to promote the technologies they had available for spin-off: e.g. field agents, demonstrations and field days, market research, professional announcement services for media and managers, etc.[10]. Too many departments, even the best, seemed convinced of the principle that publishing the availability of a technology was sufficient. One or two universities had recognised that traditional policies on personnel and scholarship would be increasingly inadequate as linkages grew. However, the recognition had not proceeded past beginning to think about such issues as criteria for tenure and promotion, temporary attachment to industry, base salary versus fee for service work, legal conflicts and the universities' responsibilities for safe and professional work.

Despite this environment, many initiatives are being taken by professors and researchers and, as seen below (Chapter 6, Section 3), a number of innovations and high technology ventures are originating from university teams. However, in most cases, they experience difficulties to transform a laboratory success into a commercial one. The absence of an adequate framework and climate for communication and collaboration with industry explains this. The need for improvement of university-industry relations is well recognised as illustrated for instance by the document *Partnership for Growth*[11] recently issued by an inter-Canadian conference of university deans, chaired by the President of the University of Regina. Actions are urgently needed, supported by government authorities, to create a climate in which those efforts can be expanded and rewarded.

b) Publicly-supported R&D in micro-electronics

All the western provinces have some form of publicly supported R&D in micro-electronics and all have plans for enhanced activity. In addition, every university we visited had a programme in micro-electronics which was presented as an indication of the university's and province's support for research in this field. Three elements which are essential if these infrastructures are to have an impact are: professional talent, state-of-the-art projects and modern equipment. In each of these three aspects, the programmes visited by the review team in Western Canada certainly pass the threshold of professional quality. The staffs are trained professionals, their equipment is up to date (but not the most advanced) and their work is at the state-of-the-art (but not pushing it into new territory). In short, the capability is there to provide professional assistance for new ventures in micro-electronics.

However, a fourth element is also necessary for an influential infrastructure: leadership which is dedicated to quality service to each province's electronics industry. Top management of infrastructure programmes must take steps such as: encouraging spin-offs of new ventures, acting as a catalyst for informal networks centred on micro-electronics, supporting promising people in provincial organisations, sharing facilities and talent with local ventures, providing advice and encouragement to entrepreneurs, etc.

The evidence gathered by the review team – on the basis of a limited exposure – on this critical element shows that leaders of research programmes need to adopt a more positive attitude. Too often the dominant orientation of top management was to "showcase" their own capabilities, rather than support others. They were primarily focused on how much status and business they received from outside the province, rather than how much leadership they provided inside the province. When we asked how many spin-offs there had been from their programmes, we were told that their work was proprietary or that "while there have been one or two, we can't encourage our best staff to leave". When we asked about research sponsored at universities, leaders instanced with pride small programmes funded at status universities in Eastern Canada or in the US. When we asked what areas of technology they were advancing, we were told about the needs of US space and defence programmes.

c)   *Publicly-supported R&D in biotechnology*

The situation in biotechnology is quite different from that in micro-electronics. In the latter there is no existing base and each province is basing its plans on impressions acquired from popular fervour about this field. In biotechnology, there is a base of existing talent and research which forms a platform of local competence to support the sophistication in science and size of programmes needed for the laws of chance to yield new ventures. The base in biotechnology is animal and plant biology and we visited important programmes in each of the provinces.

In most of the provinces, there are sophisticated forms of organisation in advanced fields of biotechnology. Universities are working closely with companies spun off by university faculty and students. In a prototype case which we encountered, the spin-off was located at the university in the dual form of a commercial venture and academic research programme on animal virology (Biostar, Inc. of Saskatoon). While the dual nature of the programme was creating identity problems for individuals and managers, it also assured close interaction between commercial work and the frontier of progress in science. The result was mutual stimulation of both the commercial and academic programmes and a steady, if small, supply of highly trained professionals.

The close ties between science and commerce in biotechnology may create problems. But these should not overshadow the promises. The present low employment stems from two factors: the capital-intensive nature of biotechnology and the apparently slow growth of western firms. Few biotechnology companies manufacture products requiring production staff. As well, the complexity of their work and need for long test and screening programmes (i.e. regulatory clearance) do not allow rapid growth. Thus while biotechnology may enhance the scientific reputation of Western Canada, it has not provided a rapid return for politicians and the public supporting innovation policies. In the long run, however, the competence that is built may yield stable strength which will slowly produce many firms in such related industries as instrumentation, food technology and materials processing.

d)   *Publicly-supported R&D in agricultural technology*

A strong research infrastructure exists in agricultural technology. On the one hand, the large programmes have the characteristics of the natural resource programmes (professional, but conservative in management – see below). On the other hand, there are small programmes which are more innovative and more exploratory in nature: e.g. Farming for the Future in Alberta, Innovative Acres in Saskatchewan, Grassroot Project in Manitoba.

However, the team found repeated reason to encourage the western provinces to further expand their efforts in extension services, especially as these relate to the need to work with industry as well as farmers. The federal extension agents we met were of high quality, but clearly too overloaded with existing tasks to aid in the development of new industry. For example, in one province a strong, federal programme had virtually no ties or orientation to provincial industry. In another case, a provincial laboratory was working on an innovative food product. While the staff had good ties to farmers who produced the raw material, they were proceeding with no input from the food industry (e.g. cost targets, taste preferences, competitive product specifications, etc.).

e) Publicly-supported R&D in natural resource industries

As the economic base for Western Canada is the natural resource industry, it is not surprising that the provinces are proud of publicly-supported R&D in their natural resource industries. Typically, such programmes are located at universities, controlled by joint boards of public and industry representatives and closely related to university educational and research programmes. (An exception is Alberta's tar sands and heavy oil projects which fund work in industry.) In every case, we again found that the staff of such programmes had sound backgrounds in both their discipline and knowledge of industry, were operating with up to date equipment and were working on projects at the frontiers of their field.

However, in no case did we find that such programmes led to the creation of new ventures or breakthrough discoveries, or acted as a catalyst for networks of opinion leaders in their industry. For example, we asked the directors of a programme if there were any innovative practices which could be traced to their work of over 20 years. The reply was that the purpose of the programme was to produce data to support the industry, not to do R&D on new technology.

The fact that natural resource programmes are acting as infrastructure support for existing technology, rather than generators of new technology, is not necessarily a weakness or an indication of poor return on public funds. The programmes are completely professional and probably do yield a public return. The issue is that innovation policies of western provinces are based on the idea that R&D can create new jobs, high-technology status and lessened dependence on commodity industries. Yet the largest allocations for R&D (by order of magnitude) are for servicing existing, capital-intensive, natural resource activities. We are thus led to discuss broader issues raised by the development of innovation programmes within economies dominated by natural resource industries.

4. Natural resource economies and innovation programmes

The history of government efforts to stimulate knowledge-intensive industries is rife with failures stemming from the belief that technological brilliance is sufficient for success. Country after country and time after time, policy makers have invested huge sums on the promise that a technology will be the basis for a new industry (e.g. solar power, supersonic airplanes, electric automobiles, etc.). Policy makers in the western provinces are all too aware of this history of governmental failure and all too committed to the belief that close co-operation with industry is the key to avoiding failures and to developing sound programmes. Throughout our visits with groups in every province, this perspective was evident in the compositions of the groups we met, the policies underlying technology investments and the attitudes of governmental officials. The difficulties arise when one defines the mix of industrial executives who will work closely with government, and when one considers the economic and environmental forces which shape the mind set of these people.

Western Canada is a world-class participant in a huge industrial sector which provides a plentiful supply of executives with wide ranging experience in practical operations and the market. The reference is to Western Canada's extensive system of natural resource industries. These executives are very active in the technological strategies of the provinces and in the overview of the specific programmes which result. The experience of these executives and the mind set which has resulted is of critical importance in the formulation and implementation of technology policies in Western Canada.

a) The mind set determined by exploitation of natural resources

Anyone hoping to operate profitably and to survive in the natural resources industry must respect certain attitudes, principles and styles of management, which are determined by the industry's two dominant features: the unpredictability of external factors and the massive scale of operations. The details of these factors are summarised below:

i) Unstable prices: natural resources are commodity products and their prices fluctuate over time bringing periods of prosperity and recession. Once one has been through one or more of these cycles, one learns to be very conservative with capital, especially during boom times;

ii) Unstable policies: the massive size of the resource industries in Western Canada attracts government attention, especially during periods of prosperity. Government fluctuation between support and taxation of profits leads to a suspicion of government and to preoccupation with the importance of lobbying and being conservative about shared plans;

iii) Unstable environment: many technologies and methods of operation which work well elsewhere have unexpected and severe breakdowns under the cold, windy and remote conditions which prevail in Canadian winters. Since it is impossible to tell if something will work in this environment until it is tried and since the trying is very expensive, executives learn to be moderate in enthusiasm about new ideas and insistent on proof of feasibility;

iv) Permanence: the natural resources of Western Canada are massive, they have been there for millions of years and will, in terms of human life spans, always be there. One can, therefore, afford to be slow and careful in their exploitation;

v) Massive scale: consistent with the scale in which the resources appear, the systems for exploiting them are also massive. Huge mining machines, conveyor belts, saw mills and transport networks which establish world records are common. The result is that when there is a failure in any one element, it can be disastrous in terms of its own cost and impact on the system. This fact of systemic amplification of failures reinforces the need for very careful planning and long experimentation. In addition, the massive scale of operations requires a perspective which envisions huge projects costing hundreds of million of dollars;

vi) Institutionalised markets: since the natural resources are all well established commodities (oil, timber, potash, etc.) the market processes for their sale and distribution have been established for some time and are relatively stable. Competitiveness in commodity industries is achieved through low production and transport costs, rather than sales and distribution.

In short, the characteristics of the natural resource industries of Western Canada emphasize strength in operation management and control over the cost of production. The executive style and attitudes which are promoted emphasize skill in careful planning of complex and detailed systems, suspicion of government plans, reliance on the tried and true and unwillingness to assume risk in the face of the high and uncontrollable risks of the surrounding environment.

b) Consequences of this mind set

The natural resource mind set contrasts sharply with the mind set which is functional in knowledge-intensive industries. That is, when initiating new firms based on R&D, the key strengths must be in marketing and R&D which stress growth through satisfying market needs better than the competition. The styles and attitudes needed for these companies emphasize willingness to assume risk, to try new and untried ideas despite the fact that one never knows what will happen, to be challenged to work with projects which cost many thousands of dollars while retaining the vision of future markets, and to work quickly to build a lead over the competition[12].

The basic concept behind industry-government co-operation in planning innovation programmes is that the plans will reflect the realities of industrial practice. This result has been achieved. However, the reliance on the type of co-operation which has developed in the natural resource sector will lead to an inappropriate result, different to that desired by the government and the public. In particular, the impact will be felt in the objectives set for programmes, the nature of the networks through which information flows and the private response to public action on technology-intensive industries.

A keynote of one province's programme for industrial innovation is that: "Oil will provide growth and agriculture will provide stability". The emphasis on natural resource industries is all too understandable when one considers the capital requirements of exploiting massive and difficult resource. In the future, this statement of goals may be reversed, so that oil provides the stability and agriculture the opportunities for growth, as the latter, in the long term, should benefit from technical breakthroughs providing important gains of added value (eventually with a reduction of the number of farmers). The perspective of large scale projects is a natural extension of experience, while it is difficult to perceive the potential for growth in existing industries which are currently small and/or less science intensive.

Officials in charge of implementing the resulting projects require continual input and guidance in order to make the decisions which determine concrete progress. This continual stream of decisions and learning works best when there is active and informal interaction among a network of people who share a common commitment to the programme and complement each other in possessing the different kinds of expertise needed. Such networks operate with a great deal of informal communication on a person-to-person basis via telephone, small group meetings, exchange of visits, etc. In short, the needs for informal interaction clash directly with the types of experience and managerial styles possessed by resource industry executives. The consequence of this clash is that innovation programme directors find that they lack the stimulation and assistance which arises from informal networking. However, while they sense something is missing, they are unable to recognise its nature and to take the extensive steps needed to build the right network. Existing networks develop slowly over time on an *ad hoc* basis and may lack the depth of experience and breadth needed for the ambitious goals which the public and programme leaders have set. Even more important, such unofficial networks lack the legitimacy and focus which come from major projects which have a concrete accomplishment as a target.

49

It is very easy to overestimate the power of government programmes to transform society even when they are massive, highly visible and personally promoted by leading figures. However, modern economies are massive and possess considerable inertia. Public funds and influence are quickly diluted in application and their specific impact turns out to be quite small. The key to success, therefore, is to use the public programmes to "leverage" private funds and create enthusiasm for the public goals. Action by private industry which supports the goal of building knowledge-intensive industry is perhaps the most important reason for co-operative planning of innovation programmes.

However, response from private industry in support of provincial goals on industrial innovation does not always reach expectations. For example, in one province we found:

- A programme funded by large amounts (hundreds of million of dollars) of private and public funds had yielded only 25 patents ("Patents aren't important to us").
- A period of 23 years used to prove the capabilities of a local firm before its mechanical equipment would begin to be purchased.
- Belief that there had been no spin-offs of any importance from a programme funded with over a billion dollars.

This is not necessarily an indication of lack of interest in innovation or commitment to Western Canada's aspirations. But one cannot expect resource industries (the major industrial powers in Western Canada) to change buying, innovation and financial practices whose necessity have made them an accepted part of everyday thinking.

The point is that such mentalities are not contained uniquely in natural resource industries. They have spawned even in industries operating in high technology sectors. For instance, in one province a private corporation holding a monopolistic situation in telecommunications behaves like a natural resource firm: no interest in spin-off, purely customer driven action, reactive R&D planning, opposition to deregulation measures which would break "concessions" obtained decades ago.

One cannot expect change in such well-established industrial attitudes and styles of operation without a major effort to understand this mind set and plan an explicit programme of commitments which transcend it. For example, the Province of Manitoba has defined an innovative contract with Canadian General Electric which has explicit goals for local spin-offs, purchasing and R&D as part of a programme to develop a major natural resource (hydroelectric power). In addition to the specific targets for local action, the programme contains explicit measures for joint monitoring by industry and public representatives of its progress. This contract is certainly a model which could inspire the whole region.

VI. STRATEGIES FOR GROWTH

The previous Chapters have provided an overview on those general conditions which should be fulfilled to create a climate conducive to innovation and a knowledge-intensive economy. However, resources are limited and the western provinces cannot be involved in every sector. The western Governments are well aware of this problem and have set up priorities for focusing their support. Choices have to be made between further support for R&D in the primary sectors (which might be very demanding from this point of view) and investment in new attractive fields (micro-electronics, biotechnology, CAD/CAM, etc.). Some provinces are emphasizing technology creation while others are focusing on technology transfer and adaptation. In both cases they are looking for appropriate market niches in North American and world competition.

This Chapter attempts to provide insights on the technical-industrial strategies to be pursued. Firstly we set out a few general principles for an overall approach to the various sectors of the economy. Secondly we discuss the problem of reaching critical masses – a condition for success in high technology. This analysis will lead to consideration on the need to build provincial technological identities. Finally some remarks are made on the decentralisation and regionalisation of innovation policies in the federal context, an important factor for technology development and economic growth.

1. General considerations

Strategies for growth, both in wealth creation and increasing employment opportunities, require clear priorities and goals. An operational framework is needed for these priorities and goals which builds on and rationalises the existing resources and facilities. There are in effect two aspects which need to be considered: the promotion of new technologies and their diffusion in the whole economy.

The new technologies are the dynamo for the knowledge-intensive economy. As in all OECD countries, micro-electronics, telecommunications and biotechnology feature largely in Western Canada. The strategy for innovation promotion and small firm development in high technology (R&D-intensive) sectors that was seen throughout the region is broadly appropriate. However more attention should be paid to the promotion of those complementary industries and services that support high technology and are lacking presently in the western provinces (see Chapter 6, Section 1b).

The dependance on natural resources should be converted into a strength and approached as a major opportunity for technological innovation and economic growth. Every effort should be made, whatever the inertia and difficulties to be encountered, to relate high technology to natural resource sectors (in the broad sense: including agriculture). On the one hand those sectors need to be "fertilized" or "revitalised" by the new technologies to maintain their competitiveness. On the other hand, a number of new technologies to be developed by Western Canada will find their uniqueness and their economic relevance only if they are integrated within the resource base. The best example of this is probably biotechnology, which is to have a major impact in agriculture, forestry, mining, etc., and which should be considered as a new resource technology[13]. As regards more particularly applications of new technology in agriculture which is world competitive and has the absolute necessity to remain so, relevant schemes have been launched (see Chapter 5, Section 3d), but they have just scratched the surface of the potential.

In addition to the application of new technologies in resource sectors, there is considerable scope for applications in the established manufacturing activities (notably in Manitoba and British Columbia where the manufacturing base is relatively larger) and in the service sectors, and particularly in those sectors under public control (health, education, etc.) for which procurement policies can create a customer base. More generally all provinces should have policies for helping in the transfer of advanced technological resources existing in the province and their adaptation to potential needs.

There is also a need in all provinces to establish active policies for technology transfer from outside into the province. Such technology transfer may help in providing complementary technologies necessary to develop innovations in the province and thus, at the same time, open markets out of the province, in Canada as well as abroad. Access to foreign technology can usefully be reinforced by appropriate agreements between Canada and foreign countries (see for instance those in place with Japan and Germany in which some of the western provinces participate).

Moreover it is necessary to actively pursue the establishment of basic infrastructures – e.g. telecommunication networks – useful for the development of a vibrant communication/information society, which is seen to structure the post-industrial era. Such networks would serve provincial needs arising from demographic and topographic structures in Western Canada and through this, further develop an export industry. In view of a tradition in social innovations based on communication technologies (e.g. for distant learning), Western Canada can become a buoyant laboratory of a worldwide significance for experimenting and developing all sorts of software and media products.

Analysis of the growth process of high technology-based communities in Western Canada shows that practically all the firms start by serving local needs. A majority of these firms are operating in fields of diffusion and application of new technologies (e.g. computer services). These firms are generally not high technology, but are prosperous and spin off many jobs. We call them "second layer industries" as they complement the high technology firms *per se*. When local firms have developed sophisticated, competitive products, they begin exporting out of the province and eventually enter foreign markets. This has been, for instance, the case in telemetry and remote sensing.

In fact, most of the international markets that western firms are approaching are highly competitive, including those of resource-related technologies. However there is one particular advantage that the western firms should not neglect: the climatic and environmental conditions prevailing in the West. These extreme conditions force them to develop extremely high quality technologies, combining both high performance and resistance to external constraints. In most cases those technologies will be produced in very limited series for tiny market niches. But there are myriads of fields and products concerned: in agriculture (e.g. frost resistant crops), mechanics (programmable robots for ice areas), etc.

2. The critical mass problem

Innovation and business creation thrive in well delimited areas where a critical mass of knowledge and entrepreneurship has been accumulated. The existence of a critical mass is evidenced by the appearance of a kind of "chain reaction process" in the development of new products and new technology-based firms. To understand this concept of critical mass a simple image may be useful. Critical mass is reached when, among a certain group of persons, two or more with complementary skills decide to join together to form a new company or develop a new project. Thus they do not limit their relationship to single exchanges of goods and services. Their decision is motivated by mutual identification of complementarity in knowledge, skills, technology, markets, etc. So there is a synergistic process leading to something new with a competitive edge. The resulting combination has an effectiveness greater than the sum of its components.

To sustain the development of a critical mass in high technology sectors, some basic conditions are required. Firstly, there is a need for strong knowledge producing infrastructures (universities, research laboratories). There is also a need for a sufficiently diversified industrial and commercial base which can fulfil technical needs of intending entrepreneurs, provide a market place, and offer possibilities for spin-offs. There is finally a need for money which should be immediately available to support intending entrepreneurs and projects. These conditions – particularly the first two – are unequally fulfilled in the western provinces. But the presence of critical masses in particular niches is clearly identifiable among the various high technology communities. The importance and nature of those niches vary in the different provinces.

In British Columbia critical mass has almost been reached in the large area of electronic systems and notably those related to telecommunications (e.g. remote terminals, signal processors, data transmission, image processors, chip design). This is evidenced by the curves representing sales and employment of the electronics industry which are climbing at a fast rate (see Figure 3). A significant number of entrepreneurs have been attracted from other provinces and the United States. The range of activities and products covered by the electronics industry in British Columbia is important. Moreover a basic industry exists providing most of the needed services in system design, facilities in quality control and assessment, and supply of components such as printed circuit boards.

In Alberta, it seems that the critical mass has been reached in certain segments of the telecommunications sector such as remote sensing, thanks in particular to the presence of Northern Telecom and its research centre (which has been unfortunately shut down in April 1986). It is also obvious that a critical mass has been reached in areas of electronic devices related to geophysical and seismic applications. This was boosted by the needs of the oil and gas sector and because of a huge concentration of specialists in geophysics – a seventh of the world geophysicists are located in Alberta.

In Saskatchewan, critical mass has been achieved in certain segments related to telecommunications and electronics, and notably the segment of electronic instrumentation and transmission devices (modems, satellite receivers, electronic tracking systems, analytical automatic products). The process has been largely based on the foundation, fifteen years ago, of the SED Corporation which has been developing a wide range of high quality electronic instrumentation devices, and from which there has been a number of spin-offs. Further efforts are needed to reach critical mass in other sectors. The most promising fields are probably those related to biotechnology, in view of the strengths of the agricultural and agricultural business sectors.

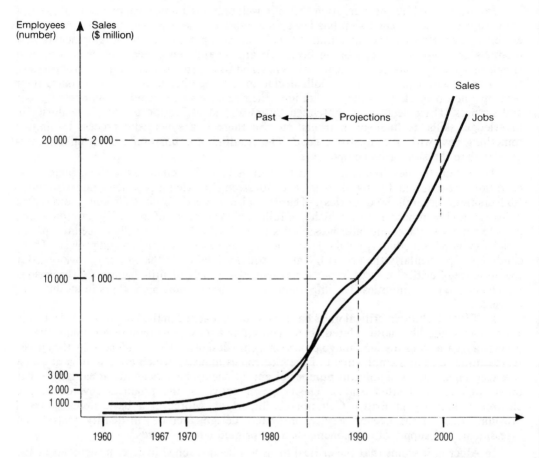

Figure 3

BRITISH COLUMBIA ELECTRONIC MANUFACTURERS

Sources: OECD Secretariat and Statistics Canada.

In Manitoba, where emphasis has been put on diffusion and adaptation of technology, the review team has not identified a critical mass in the advanced technology sectors. Future possibilities may be located in those technologies, such as CAD/CAM, related to manufacturing activities, taking advantage of the presence in the province of firms in aerospace and huge projects in nuclear industry and hydroelectricity.

To sum up and, to a certain extent, support our remarks on the situation of each province from the point of view of critical mass in high technologies, the table 15 provides figures on the level of activity of a major international (US based) computer and instrumentation company in the western provinces. This company is specialised in basic systems for industrial use and research laboratory equipment, and it supplies more than 65 per cent of the world market for

Table 15. **Level of activity in the western provinces of a major US-based computer company**

	Manitoba	Saskatchewan	Alberta	British Columbia
	Employees			
1979	17	12	24	48
1984	19[1]	0[1]	26	81
	Sales, C$ thousands[2]			
1979	1 700	1 400	3 200	4 800
1984	3 800[1]	–[1]	4 500	7 000

1. In 1984 services for Saskatchewan were handled from Winnipeg.
2. Percentage of sales: instrumentation, 64 per cent; computers, 20 per cent; medical and analytical, 16 per cent (Alberta only).

such products. Thus the level of its sales is a good indicator of the whole level of activity and dynamism of the high technology sectors in the communities under consideration.

These differences are attributable to several factors: the overall climate for business and innovation is steadily improving as one goes from east to west, the development of the high technology industries is more recent in the prairie provinces and, above all, the sizes of the economies concerned differ greatly (in terms of population, Manitoba and Saskatchewan are half of Alberta and British Columbia). This shows the importance of marshalling and focusing the efforts on strengths existing in the provinces. The key to such a focusing approach is the development of a sense of technological identity in each province.

3. Towards technological identities

The primary purpose of an identity is to focus resources and efforts by concentrating them on those fields and opportunities where there are special advantages. In the course of such focusing, each province will also develop its understanding of the resource commitments that are necessary for commercial, as opposed to laboratory, success.

Much of the "fuel" feeding the West's commitment to technology is a vision of exciting innovations which will produce another Northern Telecom, if not an IBM. Despite the fact that almost every region of every OECD nation has similar visions, such visions can provide the energy and public support needed for the long term task of building an industrial infrastructure. The key to making this energy a positive force and avoiding groundless fantasies is a leadership of specific people who have the reputations, connections and commitments necessary to marshall specific resources and focus them on competitive opportunities.

In addition to vision which is mobilising leadership in specific fields of high technology, each province possesses existing infrastructure. That is, markets, suppliers, experience and tested people which provide a comparative advantage in what we have called the second layer of industry: fields which are not necessarily high-technology, but which spin off many jobs, service companies and small, prosperous new firms. Finally, each province possesses existing strengths in transferring technology into its industry – both technology available in the province and technology from outside.

Although a short visit did not permit the team to come to definitive conclusions about each province's technological identity, we shall offer some observations to illustrate the usefulness of an identity. Specifically, we shall highlight the elements of visionary focus, infrastructure bases, and technology transfer opportunities. Since the essence of identity is a focusing of effort, and since this necessarily means constraining resources, we shall also discuss those areas where each province may be overestimating opportunities.

Manitoba

Industrial infrastructures appear to be developing strength in two areas: electric power distribution and food technology. The province's contract for hydroelectric systems deserves special mention as an impressive model for all of Western Canada (see Chapter 5, Section 4). Both electric power and food technology could easily become the "launching sites" for many new firms specialising in services, local needs and specialised niches in the markets. However, care must be taken not to overemphasize R&D in such glamorous areas as direct current transmission and control instrumentation. Such technologies are all too easily imported from other regions which are devoting greater resources to intensive R&D.

In both the fields of food and power technology, the University of Manitoba has taken an active role. However, the caution about overemphasis on R&D is especially relevant when universities are active participants. Impressive talent is there. However, as with all western universities, there is a strong need to increase competence in the managerial dimensions of technological innovation. Such competence is especially necessary in marketing (defining a market, understanding customer needs, assessing competition and organising channels of distribution) and planning a supporting infrastructure of suppliers and service organisations.

Manitoba deserves recognition for its progressive policies to encourage transfer of already developed technologies. In particular, policy makers have emphasized the field of computer aided manufacturing and design (CAD/CAM). However, the emphasis is, once again, in danger of being too broad, overly focused on R&D, and anchored on weak and public institutions. What is being relatively underemphasized is the province's emerging excellence in precision engineering at several firms. There are also emerging capabilities on both sides of suppliers and markets in biomedical engineering. Start-ups (based on university origins) are providing valuable exposure and experience in this promising field.

Manitoba pays particular attention to information technology and the Info-Tech Center, recently established, constitutes an important step. This centre seems to address the full range of issues, at least in its plans. It provides basic technical advice, legal support and financial aid to entrepreneurs. Activities have been launched with health and transportation organisations to develop procurement policies for software and computer systems. Awareness of information technology is promoted in schools. Major foreign computer and electronics companies have graciously contributed to the centre in providing basic equipment. The centre is too recent – it opened its door in the autumn 1985 – to see if its promises are being realised. To make it a success, it is important that visible results (e.g. development of new businesses or products supported by the centre) be achieved in the short term.

Saskatchewan

As mentioned above, in Saskatchewan two fields of high technology are very promising: agricultural biotechnology (e.g. animal disease diagnosis and treatment) and signal-processing electronics (e.g. instrumentation, satellite communications). Both fields are anchored in world level accomplishments and exceptional talent in science and technology.

Once again, both biotechnology and signal processing are strongly identified with university origins. The province appears to have taken constructive steps in signal processing to promote the networks needed to attract industrially experienced leadership. Similar steps are warranted in biotechnology in order to minimise the chances of a repetition of the problems of having commercial organisations run by academic norms, which plagued signal processing ventures.

In the field of agricultural technology there are impressive talents among farmers and others in applying new technology to agricultural problems (e.g. instrumentation for control of feeding processes, agricultural software) or in developing inventions based on more conventional technology (for example, the majority of exhibits at an inter-provincial farm show originated with local inventors seeking to improve current operations by using basic skills honed in machinery repair). A vital second-layer industry is developing; it should be consolidated and expanded.

There is an emerging competence in computer software. Public sector agencies (in health, education, environment, etc.) have stimulated innovative ideas such as interactive videos and educational computer networks. There is an important fiber optic network in development. Those factors taken together may provide a basis for further advance in information/communication technology.

These examples show the creativity of the Saskatchewan people. However it should be admitted that a number of innovations and firms are at an early stage of development. What is needed now is to take steps to promote them, notably by adequate exposure to national and international competition and an active technology transfer policy[14]. That might be more urgent, in the short term, than promoting new fields (e.g. CAD/CAM) not obviously related to emerging strength and infrastructure.

Alberta

Unlike the provinces described previously, the high technology fields which impressed the team in Alberta had their origin in the work of industry, rather than academia. In addition to experienced industrialists and impressive achievements in technology, this was one of the few instances during our visit where we observed the emergence of the kind of networking between multiple enterprises which is needed for long term and broad advances.

There are strong points in computer technology and in telecommunications (e.g. telephone systems). Alberta is also well on the way to establishing a strong, second layer industry in the field of data handling and processing technology (e.g. remote sensing and data handling, computer software, geophysical data processing, seismic measurement, etc.). The emerging strength has been anchored in the infrastructure surrounding the province's natural resource industry. Should the conservative networks which control natural resources recognise this strength and give it increased and early support, Alberta will have an engine of growth which could be the envy of Western Canada.

The field of data handling and processing is not only a field of promise for ventures on the fringes of high technology. It is also the primary area in which the province is active in technology transfer. If public officials work with entrepreneurial networks to develop useful infrastructure in practical frontier areas (e.g. university and other research centres on decision support systems, interactive graphics and distributed network management), rather than emphasizing futuristic areas, the province may support a vital service industry.

In the agricultural sector the review team has identified an innovative programme (Farming for the Future), a base of competence in food technology, with biotechnology entrepreneurs in such new fields as embryo transplants, and met several innovative farmers.

The point is that there has been little official encouragement, so the leadership has become scattered and too easily controlled by non-innovative academics and farm organisations. The result has been the overemphasis on basic research and improvement of existing technology. In contrast, there is an emerging network of farmers, entrepreneurs and government officials who recognise the potential and are beginning to take the steps to build programmes and policies which will develop this potential into commercially viable ventures (e.g. developing a system for market intelligence, attracting the suppliers needed for agricultural technology, staffing research centres with industrially experienced personnel, etc.).

British Columbia

The electronics industry is a strong point in British Columbia. The technological achievements, professional staff and promising products of companies are a current asset and base for realistic expectations of future growth. Should the established giants of British Columbia industry respond to this capability with an aggressive programme of joint ventures, purchasing support and promoted spin-offs, the province would be able to build the managerial networks and experience for competition on a world level. Moreover the high technology community needs better networking to exploit fully its potential, particularly to enter successfully in the highly competitive field of micro-electronics (micro-chips, super computers, etc.).

As with most of the western provinces, the natural resource industries provide an industrial infrastructure for the growth of second layer, technology-oriented industry in British Columbia. There is technological competence and entrepreneurial talent in such fields as: hydromechanics (ore slurries), forest products and the use of genetic engineering in mining (biometal winning). In addition, the field of offshore technology offers the province exciting opportunities in combining its strengths in electronics with its established infrastructure in fishing and ocean operations, notably with the Institute of Ocean Sciences, established at Victoria.

On the other hand, the natural resource industries themselves need strong reinvigoration with intense R&D and marketing effort. The forest industry is now at a cross roads and probably in a mind state where it is ready to embrace new technology. It has also recognised the urgency of increasing the value added in timber products. The newer mills are efficient and well equipped. There is little research and development carried out in British Columbia by the forestry companies. These companies also do not contract out research at the universities. Forest research at the universities is academic. Forintek Canada Corporation (a private corporation which receives funding from the Federal Government, Provincial Governments and the Canadian wood products industry) has an excellent laboratory facility and could contribute significantly to a turnaround in the forest sector. Some innovative products and processes have been developed at Forintek, but the mechanisms for commercialising them seem inadequate.

Mining in British Columbia is a major resource industry that has been severely hit by declining prices, reduced capital investment, low cash flow, and outdated technology. The result on competitivity has been catastrophic. The mining industry is locked into a low growth cycle and there is very little in-province research and development to regenerate the industry. The majority of mining companies either do no research or do their research in central Canada. These companies also appear to contract little research to British Columbia universities.

Of all the provinces, British Columbia has the strongest conviction in the laisser-faire approach to public policy on technological innovation. One consequence of this conviction has been, so far, a limited policy support for technology transfer. Recent initiatives have been taken for promoting technology transfer both from inside the province (e.g. from university) and from outside. These initiatives should be strengthened.

4. Regionalisation of innovation policy

To conclude this analysis it is worthwhile to say one or two words on the regionalisation of innovation policy within the federal context, a process which is well underway with the Economic Regional Development Agreements and related sub-agreements in science and technology, and which is important for designing appropriate strategies both at regional and national level.

The central fact about the regionalisation of innovation policy is that each province has a unique cultural and economic context for innovation. Basic differences exist in economic resources (e.g. the oil of Alberta and the hydro power of Manitoba), cultural heritage (e.g. the Pacific Coast mentality of British Columbia), politics (e.g. a long history of political independence in Saskatchewan) and especially in leadership (e.g. the chief advocate of science and technology in Alberta was the Premier, Saskatchewan has charismatic leaders in charge of science and technology, Manitoba has well-established industrial leadership in basic industries and British Columbia's policy was strongly influenced by the personality of a scientist minister).

The result is different strengths in infrastructure, different power structures, different resources available for investment, and different goals and choices of fields for development. Given these differences, the regionalisation of innovation policy appears an inescapable choice for major programmes and policies. An especially important factor which reinforces this choice is that the success of innovation policy depends on the co-operation of leaders in established industry, academia and the entrepreneurial community. Such co-operation and active interaction is only possible if policy makers are known and respected personalities in active communication on a personal level and known to be personally committed to living and prospering in the province. In short, the review team found that the argument that regionalisation of policy led to greater responsiveness was supported repeatedly throughout the review study.

While the review team found repeated evidence for the responsiveness of regional policy, it did not find support for the idea that regional programmes were less costly than federal programmes. While regional programmes were less costly on a province-by-province basis, it was not evident that the total provincial staff across Western Canada were fewer in number or that provincial officials were more productive than federal officials. In other words, it appears that little cost saving is possible when the total effort expended across the West is considered.

Instead of enhanced productivity, the review team did note the persistent presence of administrative overload. That is, regional staff was small and though the size of the programmes (in dollars) was smaller than the federal ones, the breadth of the programmes in variety and number of actions was comparable. The result was that regional officials are short-staffed, must continually switch from issue to issue and lack the time and funds for in-depth study before decisions are reached. While such working conditions are found in many federal agencies as well as regional, the review team found the degree to be significantly greater at the regional level. In addition, federal officials possess significantly higher budgets to contract out the consulting and expert studies which provide essential policy analyses valuable in carefully deliberated and innovative initiatives.

Moreover there is a situation of extremely fierce competition between the provinces. This encourages a wasteful duplication when competition is not for the best programme structure but for federal dollars. The primary cause of potential waste stems from the fact that a significant part of the drive behind policy in each of the provinces is the same "high tech myth" found throughout the world: namely, that the brightest hopes for the future lie in such fields as micro-electronics, opto-electronics and CAD/CAM. As a result, each province pursues dreams of winning a federal grant for *the* West's most important centre in at least one of these fields. The waste, especially in the opportunity costs of effort deflected from more realistic and promising directions, can be significant.

Interestingly enough, the review team found that, when the provinces were pursuing the same directions and when these directions were anchored in fields where each province had an existing infrastructure and leadership, there seemed to be so much potential for practical progress that no waste (or competitive stimulation) was evident. This observation especially applies to the fields of food processing and agriculturally based biotechnology which are being actively pursued in the three prairie provinces.

Finally this situation of competition does not encourage co-operative efforts between provinces, while they would benefit greatly from search of complementarities and exchanges of experiences. Each of the provinces has much to learn from the other ones. From this point of view, this report is a significant step forward in the right direction.

VII. POLICY RECOMMENDATIONS

Our policy recommendations will be presented as follows: the three first headings will summarise the main conclusions outcoming from the previous Chapters and which concern the development of a technological identity (Section 1), of an environment more supportive for new industry (Section 2), and of human resources (Section 3). Then we shall discuss a theme which has appeared repeatedly throughout the report: the need to multiply structures networking people and organisations into a critical mass (Section 4). Proposals will then be made for inter-provincial co-operation (Section 5) and collaboration with the Federal Government (Section 6). Finally some specific measures will be proposed for each province (Section 7). In all cases we propose a few key recommendations rather than an exhaustive list. It may appear also that some provinces have already implemented certain actions that we suggest. Their experience would be most useful for other provinces.

However before exposing these recommendations, we would like to emphasize some basic characteristics of innovation policy. It should be conceived as a new policy concept (or tool) and not as a new policy operating on a specific administrative territory with well delimited frontiers. It is, in its very nature, an *inter-departmental policy* operating horizontally through the established, vertically structured, government departments: science, technology, industry, trade, energy, agriculture, finance, education, labour, etc. *The effectiveness of innovation policy will depend upon the extent to which policies implemented in all spheres by these ministries are called into question.*

To accomplish this task, appropriate mechanisms need to be established in Governments. Ministers responsible for science and technology could accomplish it, providing that they have the possibility and status to operate as inter-departmental co-ordinators with full freedom. An alternate approach would be the establishment of a *Special Commissioner for Innovation*. This would be a fixed term appointment, preferably of a non-public servant experienced in innovation and entrepreneurship, to be located in the Office of the Premier and thus free of departmental constraints. It would be responsible for execution of a provincial innovation policy, inclusive of setting priorities, co-ordination of innovation related activities of all relevant departments, and also orientation of innovation related activities of the Crown Corporations. The establishing of such a Commissioner could be also a means, in certain provinces, to re-mobilise the creative capabilities and propel energies towards new directions.

1. Development of a technological identity

Recognising that the strategies of the western provinces emphasize high technology sectors which are the same target areas for most national innovation programmes in OECD countries, we recommend:

- *As regards basic infrastructures of interest for the whole region, to invest in those technologies which are clearly related to its needs and strengths*: e.g. *telecommunication networks* in view of their importance in modern economic development, and of advances acquired by the western industry; and *biotechnology facilities* in view of impacts of biotechnology on the resource base of Western Canada and competences already acquired by western universities and industries;

- *As regards provincial strategies, to focus resources and efforts by concentrating on those fields and opportunities where there are special advantages*: research and development capacity, markets and suppliers providing the necessary infrastructures, human talents in technology and management, and leadership able to marshall and focus resources. Suggestions have been put forward on those fields which seem particularly promising in each province, and on which it can build its own technological identity (Chapter 6, Section 3);
- *To expand strategies and programmes for enriching the existing industry by adequate diffusion and adaptation of new technologies* (e.g. procurement policies in service sectors under public control such as health and education; further expansion of extension services in agriculture, etc.);
- *To pursue active policies for technology transfer from outside into the province*;
- *To emphasize development of high quality products*: taking notably advantage of the climatic conditions prevailing in Western Canada which force to produce highly performing, resistant and durable technologies.

In essence, the strategy for growth is to promote new technologies and to integrate them in the whole economy. That will put in place a *relevant* high technology sector firmly *rooted in the strengths* of the western provinces. Such a high technology sector would not be a copy of the Japanese or German industry but would be stamped with the specific characteristics of Western Canada. Such an "identity" would have positive marketing features on the world stage.

2. Creation of an environment supporting the new industry

Having based its development on the exploitation of natural wealths, Western Canada has a limited industrial tradition. This constitutes an advantage as far as the conversion problems experienced by regions and countries with a heavy industrial heritage are minimised. But this presents also some problems to the extent that the new industries are lacking adequate infrastructures and are insufficiently supported by actively involved networks in the industrial community.

A series of actions are recommended along several axes:

- *Mobilise the natural resource industries to support new entrepreneurial industries.* This requires major programmes including, for instance, promotion of joint ventures and spin-offs, subcontracting activities, mobilisation of commercial networks serving the natural resource trades. It is recommended that opportunities existing in each province for such programmes be systematically studied. This is an important aspect of policies to develop provincial technological identities, as described above;
- *Increase public support to "infrastructure" industries and services.* Specialised components manufacturers, prototype designers and builders, providers of tests and technical services, consultants in management, finance, market research, etc., are essential for the development of a high technology industry and are lacking in the western provinces. Firms in these areas are not very attractive to usual financial backers (e.g. venture capitalists) and should receive appropriate public attention and support. Government incentives (in form of subsidies or fiscal relief) could also be

provided to stimulate interaction, networking and partnership between firms already established to get full advantage of their technical and marketing capabilities. Schemes subsidising first risk purchasers could also be established in all the provinces, following recent experience of Saskatchewan. Moreover, the public sector should take the initiative with a view to establishing basic technical infrastructures and services, such as quality control and assurance, which are presently inadequate in the western provinces;

- *Improve the financial environment.* Measures have already been taken to increase the provision of venture capital. These have been welcomed and are efficient. However there is a need to establish exit mechanisms in form of secondary markets or junior boards in a Western Stock Exchange properly oriented towards the new industries, which is not the case of those currently operating in British Columbia and in Alberta. The Alberta Stock Exchange would constitute, in the view of the review team, the most suitable platform;

- *Improve the regulatory framework.* Although the review team has not examined in detail obstacles created to innovation by existing regulations, it has identified a number of regulatory rules affecting the economic activity. A large part of those rules have been imported from elsewhere where the scale and complexity of industry justified more careful regulation that it is appropriate for the West. The regulatory framework is pervasive in many areas: purchasing, hiring of employees, transportation, application for grants and contracts, required reports to be submitted to the government (on financial performance, employment situation of companies, etc.). Moreover some fields like biotechnology or drugs, important for the future of the western provinces, are subject to important and troublesome regulatory procedures and delays in provinces which lack experience and large staff dedicated to regulatory clearance. Although the review team is aware of some past or recent efforts made in certain provinces in this matter, it recommends that each province undertake a carefull and systematic examination of the current regulatory systems, to identify rules which may hinder innovation and take steps to eliminate those which fall in its jurisdiction;

- *Encourage greater university-industry collaboration.* The type of development experienced to date by the western provinces has not stimulated such collaboration. Initiatives have been taken to improve it. Further progress is however still necessary. Government authorities should design strong incentives to increase university research funded by industry (which – amounting to some 0.5 per cent – is currently low compared to international standards) as well as to promote better working procedures in the universities for diffusion of research results, patenting of inventions, setting up of teams to promote research activities in the market place, etc.

3. Enrichment of the human resources

No country can hope to build up a strong economy without considerable efforts in education and training at all levels. For Western Canada, *it is strongly recommended that both the public and private sectors make further financial efforts to sustain and expand the infrastructure in education, and in the universities in particular (need of advanced equipment, faculty staff internationally experienced, etc.).* Moreover, a number of issues have been identified. More systematic efforts are needed in several fields.

As regards *higher education* and the supply of highly qualified manpower it is recommended that each province review its need for graduates and technicians in all specialities for the next ten years. This should be based on the identification of industry's needs in each province, through appropriate surveys and forecasts and on provision of information by the Federal Government on graduates outputs at national level. Moreover there should be a systematic integration of management programmes into the technical and engineering courses throughout the whole education system, including the community colleges where students should receive basics in business management. Universities should also mount significant exchanges programmes at teacher and student levels with eastern Canadian universities and foreign institutions.

As regards *training and recurrent education*, the development of programmes of business management in the form of summer sessions or evening courses is a priority area particularly for small firms and individual entrepreneurs. Universities and technical institutes should also organise, on a regular basis, updating sessions in the form of conferences on relevant technological subjects where both the education and business communities can be informed on the state of the art and be kept abreast with the most recent advances. It is recommended that labour be involved in the preparation and running of those activities. In all these matters, Governments can act as a catalyst, and ensure that the programmes are of sufficient scale and practicality.

As regards the *management of existing human resources*:

— Conditions affecting the *mobility* of people, notably between industry and university, should be improved. Obstacles of an administrative or financial character which may prevent such a mobility need to be identified and eliminated. This may offer, in particular, opportunities to improve leadership of R&D programmes in both universities and industry;

— Active programmes should be organised by each province to locate and track the success of "native sons and daughters" *and recruit them back* once they have accumulated international exposure in technology, marketing, finance and management;

— In order to fill unavoidable manpower gaps which will appear in the high technology communities in the short and medium term, well targeted publicity campaigns should be launched jointly by the Government and industry to *attract* foreign specialists (for instance in business magazines with a world wide coverage);

— Important *exposure* measures should be taken particularly for innovators and entrepreneurs of the western provinces, and supports for visits abroad and participation to international fairs should be strengthened. Organisation of international workshops in each province is also recommended.

Finally, within this *whole effort of mobilisation of human resources*, it is recommended that awareness *campaigns* be launched in three directions:

— *To improve industrial relations* between the social partners in publicising all across Western Canada examples of collaboration leading to significant results: e.g. establishment of the Workplace Innovation Center in Manitoba, development of training programmes co-designed by the labour and the business sectors as in Saskatchewan, use of pension funds for venture capital, etc.;

— *To encourage women's careers* in knowledge-intensive industry. Specifically, we recommend a Regional Conference on Women and Technological Entrepreneurship and the establishment of "Women's Policy" committees in each province's Ministries of Industry, Science and Technology, and Education (as these matters need to be particularly considered at the school level);

- *To stimulate ethnic blending* which is a strength of (Western) Canadians, in highlighting innovations resulting from groups integrating people of different origins (European, Asian, Inuits, etc.).

For all these actions it is recommended that media (TV, radio, newspapers) be involved for broad coverage of related events and campaigns.

4. Need for networking people and organisations

The need for better networking of people and organisations under stronger leadership has been stressed throughout the whole report. We recommend that the number and variety of supportive organisations for entrepreneurs and innovators be increased. There is great scope in each province for multiplying for instance:

- *Incubator spaces.* These reduce considerably the rate of mortality of new technology-based firms. A comprehensive programme of incubators would also contribute significantly to job creation.
- *Evaluations centres*, eventually coupled with incubators, which are of great help for assessing and packaging innovative projects. It is recommended that these structures be established in university campuses. Each university should consider the possibility of opening one evaluation centre. They need to be extensively networked with surrounding expertise in technology, finance, marketing and entrepreneurial communities. It is also recommended that these evaluation centres focus on those fields which are constitutive of technological specialities of each province (see Chapter 6, Section 3).
- *Technological associations* which give the opportunity for people with converging interest to network and marshall their resources and competences into a critical mass.

More specifically for the agriculture sector, which is of a paramount importance for all four provinces, it is recommended that:

- In each of the four provinces one or several agricultural business fora be created, focused on innovation in agricultural industries, anchored with specific facilities for R&D or venture support, and representing farmers, processors, Government and academia. Such forums can do much to: increase the visibility of the specific technology as an engine of provincial growth (underestimated for agriculture); provide a model of networking between resource-based activities and high technology; promote increased access to international expertise; and assure the leverage of federal, provincial and private efforts. In all provinces, the base for such co-operative approach is present.
- At least one, and eventually several, *biotechnology parks* be opened in each province. The review team was especially impressed by the promise of biotechnology for Western Canada. The capability of the technical people met was of world class standard, the universities are enthusiastic in their role of providers of scientific breakthroughs and qualified personnel and the West possesses an existing infrastructure of depth and sophistication in agriculture. What is missing is a programmatic focus to attract international recognition, venture capital and managerial talent.

65

A biotechnology park would have many of the characteristics of a specialised industrial park (specialised offices, laboratories, centralised technical services, public relation campaigns to promote identity and attract support, etc.). In addition, we would recommend facilities such as animal holding and transport facilities, government and university laboratories, information centres, testing and data-base services, incubator space for biotechnology start ups, university/government sponsored conferences featuring internationally recognised experts, offices of suppliers who have been especially recruited for their relevance to biotechnology, etc. In addition, we recommend the location of extension offices at the park with specialists who are talented at both grassroots education and building the networks of farmers, ranchers and industrial people needed for the word-of-mouth promotion of technology and for its practical adaptation to the needs of users. The park would support work of these extension specialists with facilities for such activities as: conferencing and training, computer networking, practical demonstration and testing, "desk-top" publishing of newsletters and advisory reports, and administrative support on the communications of opinion leaders and early adopters.

It is recommended that managers of all these structures (incubator spaces, evaluation centres, technological associations, agricultural business fora, biotechnology parks) *be fully involved in the elaboration of provincial innovation policy, as full members of the "guiding council"*. The cost of establishment and operation of these structures should be largely financed by the private sector.

5. Extension of inter-provincial co-operation

The guideline for the review team was to treat the four provinces as a unit. It follows that the review team was aware of, and looking for, potential areas of collaboration that could reinforce the unity. It was recognised, however, that the provinces are very different and that independence of action was stronger than collaboration. Nonetheless some areas did become clearer as areas where co-operation would be beneficial. The following recommendations are made as tentative first steps on what is acknowledged as a difficult journey.

Exchange of experiences

Each province is rapidly moving up the learning curve for an innovation policy, and each has definite strengths and weaknesses. Thus far the co-operation between the provinces has been limited to a very few collaborative research projects, and to the group of officials which sponsored the visit by the OECD review team. It is felt that all provinces would benefit from an exchange of experience that could most effectively be achieved by instituting arrangements for regular meetings (let us say one day every two months) and exchanges of staff between agencies. The exchanges need only be short term, say two to six months, and could occur at all levels of the admittedly small agencies.

Inter-provincial co-operative projects

While it does not appear possible nor necessarily advisable to set up inter-provincial projects to be fully defined and funded by government agencies, it might be useful to encourage co-operative approaches between firms. In several European countries there are such co-operation schemes in new technology. An essential element of such co-operation is that projects be carried out by at least two companies in different jurisdictions for example, the ALVEY project in the United Kingdom, the ESPRIT programme in the EEC. The long term objective is to strengthen the technology base and obtain synergy in development. These objectives are valid for the western provinces, and a similar approach is suggested.

In certain priority fields of new technology, we recommend that a programme be set-up and jointly funded by the provinces. The operative stage should be pre-market and cover research, development, prototype building, system proving and market research. An approved project should have a minimum of two participants – organisations, companies, institutes – from two different provinces. An underlying objective is to create a network of interested bodies across the whole region, and to gain complementarity of action in the provinces. Government funding would amount to fifty per cent of the total project cost, and would match equivalent money from the private sector. Clearly, such a programme, at a later stage when its feasibility would have been proven, could be expanded to include participants from other parts of Canada and eventually from foreign countries.

Technology monitoring network

To help in inserting western technology in the international technology trends, we recommend the establishment of a computerised interactive network throughout the western provinces to follow and assess technological developments both in Canada and abroad in those fields of particular interest for the western industry. Nodal points of such interactive data bank should be established in university departments with a well-recognised expertise in relevant fields and should associate both engineering and management professors and students, with a view to develop international marketing competences. This "technology monitoring network" would usefully support evaluation centres mentioned above.

Deregulation study groups

To complement deregulation efforts to be undertaken in each province, there are several areas which would deserve a concerted approach between the provinces: agriculture with crops' licensing procedures hindering development of new seed varieties, telecommunications with standards under monopoly of different Crown Corporations in the various jurisdictions, and finance with a view to establishing a western stock exchange. Specialised groups could be formed to study relevant problems and contribute to solving them within a national context. It might also be useful that the western provinces examine, in common, international trade related regulations which may affect their activities in several fields and make appropriate proposals to the Federal Government, in relation with new trade issues (for example with the United States, or under GATT).

A Western Technology Expo

In order to raise general awareness for innovation, promote the image of western technology with its unique features, and give self confidence to the western provinces' people as a whole in their creative capacities, it is recommended that a joint exposition be organised, co-financed by the four provinces. It should have an itinerant character, staying for several weeks in each province. In the longer term it could be presented in the rest of the country and eventually abroad.

A Western Canadian Open University

To conclude this series, we suggest a major project. The provinces have already taken important initiatives in distant learning education (e.g. the Athabasca University in Alberta, the Distant Learning Authority in British Columbia). Experiences of other countries show that an open university needs to be sufficiently large to function efficiently and benefit from economies of scale. It is thus recommended that a *multipolar* inter-provincial open university be established on the basis of the experience already accumulated. It should be connected to all existing (conventional) establishments in education (notably to take advantage of available materials, e.g. video cassettes) and an overall plan of equipment (e.g. interactive computers, satellite receivers, etc.) should be designed in parallel.

6. Improvement of the relations with the Federal Government

The relations between the Federal Government and the provinces are complex and ever changing as each explores new oppportunities for growth and progress. There are many areas of overlapping competence and, for new fields like innovation policy, there are few historical precedents to follow. Despite some difficulties in the past, a new era of co-operation, mutual reinforcement of effort and sharing of benefits is well underway. It is characterised by the Memoranda of Understanding, and the ERDA agreements following them, that are currently being negotiated and brought into play.

a) Decentralisation of innovation policy

Regionalisation of innovation policy is a general trend among OECD countries. It is an inescapable evolution in Canada due to the uniqueness of each province and the federal structure. Our analysis of the regionalisation of innovation policy argued that this was an efficient approach to stimulate technological development in Western Canada. However to increase their leverage on provincial policies, we recommend that federal authorities:

- Provide major funds for programmes and support in recruiting top leaders for infrastructure programmes, but only if provincial plans clearly relate to the technological identity of each province and are strongly supported by a network of entrepreneurs, industrially experienced academics, private capital providers and public leaders;
- Concentrate funding on those programmes and projects which have strong private sector participation and which can ensure strong leadership, and sharply limit funding on those which do not satisfy these conditions;
- Strengthen the capability of IRAP officials (via added human and financial resources) and ability to relate to the major federal ministries and agencies (not just the National Research Council) to expand their contributions for liaison with regional R&D and innovation;
- Take steps to ensure that money available for procurement of goods and services – primarily by the Department of Supply and Services – is used in the most efficient way in view of the competences and possibilities existing in the West. A first approach for this would be to review in each province local offices of the Department which should publicise procurement programmes, facilitate contracts with firms and streamline tendering procedures;
- Make the best use of federally sponsored infrastructures to meet provincial needs. There are across the West excellent facilities equipped to the highest standards and staffed by first class professionals (e.g. Forintek, the Plant Biotechnology Institute, POS, the Canadian Food Products Development Center in the agricultural food area). However, these facilities need to be networked between themselves and with the corresponding provincial organisations.

b) International programmes

The insertion of the western provinces' technology in the international scene would benefit from joint approaches between the federal and provincial levels. Four aspects would deserve particular attention:

- Commissioning of "competitive analyses" (technological, market, financial and human resource data and analyses needed to understand the resource requirements for, and the managerial process of, commercial survival) in areas of technology which are commonly the focus of regional plans for knowledge-intensive economies (i.e. speaking tours by industrially proven experts, working conferences which build networks between and in provinces, studies of successes and failures in OECD nations, etc.);
- Exchanges of researchers and scientists with foreign countries (an agreement for such a purpose has been signed by Quebec with France in the field of biotechnology);
- Commercialisation of new technologies produced in the western provinces, by mobilisation of the scientific and commercial attachés networks, and programmes for international technology transfer;
- Integration in international networks of expertise and exchanges of experiences on Science Parks, Technopolises, Business Centers, which are developing in Europe, Japan, etc.

c) Provision of information

Finally some words need to be said about the *information base necessary to monitor properly a knowledge-intensive economy*. The review team was faced with a huge dossier of background information on Canada and the individual western provinces. In using this background to isolate the relevant data on innovation and new technology, severe limitations became apparent. The broad statistical base in Canada is among the best of OECD Member countries; however, it falls short in the specific area under examination. A key issue is that relevant data are generated from both provincial and federal sources, and definitions and cut off points may be different in different jurisdictions. Finally, there is the general point that some newer industrial sectors are still inadequately covered by the existant classification systems, e.g. computer software companies. More generally there is a need to document better this new economic area developing at the interface of the manufacturing and service sectors, which is of a growing importance in modern economies.

Consideration should be given, on a joint provincial-federal basis, to the generation of statistical data relevant to the new technological era. This series would include data on mainly three aspects:

- The entrepreneurial dynamism: new technology companies (numbers, establishments, liquidations), their activities, employees, needs in personnel; financial factors such as venture capital; export values;
- The investments in knowledge: research and development expenditures, manpower statistics relating to qualified scientists, engineers, technicians and managers;
- The contributions of advanced technology sectors to national (and provincial) product, revenue, employment, etc.

Those data could be published yearly. The western provinces would serve as a pilot area, while this exercise would be extended to Canada as a whole in the medium term.

7. Specific provincial measures

In the following section specific policy issues are identified for each of the western provinces. These have been selected fundamentally on the basis of our perception of the strengths and weaknesses of the individual provinces and of the most fertile areas for future growth. These policy recommendations are not designed to be comprehensive and all embracing, but to be significant steps towards a knowledge-intensive economy. While a particular province may have been chosen as a flag-bearer, a number of these policy measures have wider relevance for all four western provinces.

a) Manitoba

In Manitoba there is a need to revitalise the existing traditional industrial base and to complement existing activity with new technology. In a short time frame, some well-targeted actions are needed to give to the province self-confidence to shift resources away from traditional industry towards the new technologies and the future employment prospects that they offer.

Quality Assurance Center

Quality control and assurance facilities are lacking throughout the western provinces. Manitoba has the industrial infrastructure, the technology agencies and the manpower in place to consolidate their efforts in a Quality Assurance Center located in Winnipeg. It is recommended that such a Quality Assurance Center be set up to offer services to advanced industries in the province as well as to other western provinces. The centre could be located downtown and have high public visibility and profile. The activities of the centre would tie in well with the Technology Commercialisation Center and Program, and the Manitoba Research Council, as well as with the activities of commercial bodies like the Electronics Industry Association of Manitoba or the Small Business Association.

Opportunities for subcontractors

With its comparatively well-developed traditional industrial base, Manitoba presents more opportunities for subcontractors than other western provinces. However, most suppliers and subcontractors are presently out-of-province. A survey should be undertaken to identify products and services, required by larger firms, that could be made or provided in-province. An exposition could then be mounted under a slogan such as "We can make it" to enable local industrialists develop and show their products/services. The larger employers should be encouraged to examine the extent to which they could source locally. It may be necessary to develop a first-time-purchaser risk assurance scheme, perhaps funded by the Manitoba Jobs Fund. Such a programme should also promote "spin-off" from larger firms.

Biomedical engineering

It may be important to promote a particular sector with high growth potential that would have a demonstrative effect for the whole province and implications for a number of related industries. In Manitoba, several components necessary for an effective biomedical engineering industry are already in place. These include: research and clinical facilities at the hospital in St-Boniface; experience in instrumentation devices, in metal component manufacture and associated CAD as a result of the presence of the aerospace industry. Two other developments could also interact with a biomedical engineering capability: namely the

70

application of fiber-optics and the linkage to information technology and the use of computers. The social environment in Manitoba is particularly concerned with the needs of handicapped and disabled people which are requiring innovative and high value added devices. It is suggested, therefore, that the possibilities for establishing a biomedical engineering industry be examined. Such an industry would be able to serve the whole of Western Canada as well as some of the plain States of the US. The public sector would be a major purchaser from the industry via health departments, labour employment departments, social services and veterans and other hospitals. Procurement policies could be deployed in favour of the new industry.

b)   *Saskatchewan*

In Saskatchewan the review team was impressed by the dynamism and the self-confidence of the people. To consolidate the results already achieved, the following actions are suggested: operations to accelerate reaching a critical mass in several fields, critical assessment and promotion of local innovation by international exposure, and mobilisation of the Crown Corporations.

Critical mass operations

– Biotechnology: in view of existing talents, a highly entrepreneurial farm community, excellent infrastructures (e.g. The Protein, Oil and Starch Pilot Plant – POS, the Plant Biotechnology Institute, university faculties in veterinary sciences and virology), and actions already undertaken jointly with the Federal Government, it is recommended that Saskatchewan take the initiative of establishing a biotechnology park on the model described above. To lead this park we would recommend the appointment of at least one experienced manager who would be a consultant (and perhaps a board member) for non competing firms located at the park; the park would fit well in the present Innovation Place in Saskatoon;
– Instrumentation: to take advantage of high technical competences acquired by Saskatchewan firms in this field and of infrastructures such as the Center for Advanced Instrumentation, it is recommended that a repair centre for electronic instrumentation equipment be established, that would serve for Western Canada as a whole;
– Software: varied software products for application in health, education, agriculture, management, seem to have appeared in the recent years in Saskatchewan, it might be worthwhile to establish an evaluation centre in this field which would assess and promote such products.

International workshops

To bring a wider market orientation for Saskatchewan technology products and to better inform local entrepreneurs on technology advances elsewhere, it would be appropriate for Saskatchewan to promote a series of small (100 participants) workshops on technological topics and fields of interest to the province. Such workshops would comprise product/service expositions accompanied by lectures and discussion of the surrounding technical fields. This would serve to bring international trends to the notice of Saskatchewan businesses as well as to bring Saskatchewan products to the notice of potential buyers. Workshops should aim for 50/50 international/provincial participation and a 25/75 public sector/private sector attendance. A priority list of workshop areas should correspond to the province's technology priorities.

The role of Crown Corporations

The sense of cohesiveness which characterises Saskatchewan's society leads us to suggest that the province can undertake a pilot action, demonstrative for Western Canada, as regards Crown Corporations to promote innovative firms and their products. An analysis is required to identify how the Crown Corporations can better use their resources with a view to encouraging spin-offs, to helping new companies supplying services under long-term contracts, to promoting new products by acting as first buyer, and even to funding innovation in their spheres of interest. Set-aside quotas reserving funds for local suppliers and contract evaluation teams within the Crown Corporations would be other expected outcomes from the analysis.

c)   Alberta

In Alberta the review team was alerted to the existence of a vast body of technology that has not been fully exploited, as well as to a pool of untapped entrepreneurial talents. Moreover, with the difficult economic situation experienced with the decline of oil prices, it becomes important to make the best use of the R&D infrastructure established during the last decade and to propel energies towards new directions.

Stimulating the agricultural business

The agricultural population constitutes a reservoir of innovative and entrepreneurial talent, which could be stimulated by a concerted campaign. This campaign would include a training programme in entrepreneurship and technology, and the development of advisory counselling services. A crash programme aimed at getting a computer and modem in every farm may also be appropriate. Such a step would promote a software industry and allow the establishment of a computer-linked information system. The financing of farm innovation in Alberta should be accomplished by several related actions: mobilisation of funds in the Credit Unions, provision of funds from the producer boards, such as the wheat pool, and a five-fold increase in the funding of the programme "Farming for the Future". It is also suggested to re-implement a policy of geographic decentralisation (specifically locating agricultural centres and leaders throughout the province) which has been valuable in the past. These actions, although primarily provincially delivered and funded should be co-ordinated with the efforts of Agriculture and Environment Canada and with federal programmes such as IRAP and the ERDA agreements that relate to relevant technology. Of course, the establishment of agricultural business fora mentioned above is highly recommended. As a first task, they could assess conditions of implementation of above proposals.

Technology survey in the hydrocarbon industries

The amount of funding that has been applied to hydrocarbon resources in Alberta is considerable. Innovative outcomes of this effort should be increased in terms of patents, spin-offs and marketed technologies. Executives and managers of the hydrocarbon industries should be more aware of the contribution that they can make to innovation in the province. It is suggested that an inventory/evaluation be made of the technological content of the developed products, services and systems. This review should be made by a small expert group with international knowledge and experience in similar technology fields elsewhere. The objective of this review would be to identify latent technology, to stimulate commercialisation of this technology, and to create another wave of new technology-based firms. A survey could be followed by arrangements for funding exploitation of the technology by the Heritage Fund or one of its sub-organs. As the technology would be "Made in Alberta" it would have a significant impact on the "technology image" of the province.

Concerted research projects

With the aim of mobilising the various components of the research community, it is recommended to launch a programme of concerted projects jointly funded on a fifty-fifty basis by the Provincial Government and the industry, and associating research teams from university and/or public laboratories in several areas where there are particular strengths and needs: for example applications of advanced technologies to natural resource industry (such as robots operating under extreme conditions, expert systems for energy exploration), and development of information and computer technology in practical frontier area (such as information network management, interactive graphics). It is recommended that, through such projects, be developed new methods of collaboration between industry and university (as regards for instance exchanges of staff, diffusion of research results, patents versus publication) which could have a demonstrative value for the whole region. The projects should be sufficiently numerous and diversified in scope to have significant impact on the whole research community of the province.

d)   British Columbia

British Columbia presents a contrasting picture. On the one hand the climate for entrepreneurship and innovation in new technologies is good and few additional measures seem needed to consolidate it. On the other hand urgent actions are needed to reinvigorate the natural resource sectors and to enlarge the social foundations of innovation policy.

Electronics association

As noted in the main body of the report, the high technology community has almost reached critical mass in the electronic field with specialities in telecommunications. In order to take full advantage of such a situation, it is recommended that the communication process within this community be broadened. A new focal group can serve as a seedbed and nurture spin-offs. We suggest that the various electronics research people in the universities, in private industry including British Columbia Tel. subsidiaries, and in public sector laboratories, join together to form an association. The association would have personal, rather than corporate members, receive some provincial government support, and have as its objective the pooling of expertise leading to formation of new projects and new firms.

Natural resource skill groups

The second area concerns the natural resource branches of forestry and mining. The critical problem to be solved here is that both industries have moved, particularly from a scale point of view, from needing a particular machine/process to requiring a complete system for a mine or mill. A system, by definition needs many complementary skills. We suggest that a Forest Skill Group and a Mining Skill Group be set up in British Columbia. These groups would have the objective of pooling the requisite skills – marketing, engineering, management, research, development, software and so on – so that a critical mass able to support innovation is created. Given the parlous state of the two industrial sectors, some Provincial Government aid to the skill groups would be needed in at least the early stages. In both cases there are organisations (Forintek and/or the Council of Forest Industries and the Mining Association) which could serve as nucleus for the development of the skill groups. However, these groups would necessarily be broader and include competences of all entities interested in industry and university.

Social action

In British Columbia, there is a current of awareness rooted in social groups, such as innovative educators, women leaders and labour unions, which encapsulated their views in the recently published *People's Report*[15]. The importance of these groups is that they provide a "bottom up" view balancing the top down approach usually noted in innovation and technology policy. This refreshing new current of thinking has been particularly noticeable in the labour unions who are encouraging self-employed timbermen to bid for forest quotas, and are ready to use pension fund and credit union funds for venture capital investments. The engagement of women is also significant, and in contrast to the other western provinces. It is recommended that these networks of innovation-aware groups be maintained and developed to improve the social foundation for provincial innovation and deepen the sense of community. Regular meetings between these groups and the relevant government departments could be set up to study and select further innovative projects to be promoted on a cost sharing basis.

*

* *

These suggestions constitute a set of policy measures for immediate implementation which would serve as tests of the ability of the western provinces to overcome those obstacles which currently prevent the mobilisation of the considerable amount of talents and resources which are available. There appears in fact no limit to the possibilities of expansion of innovation and entrepreneurship in Western Canada. The transformation from the present situation to a vibrant knowledge intensive society that is tantalisingly within grasp is well underway.

NOTES AND REFERENCES

1. See *Science and Technology Indicators*, No. 2, OECD, Paris, 1986. According to the OECD definition, high technology sectors are those which spend more than 4.5 per cent of their turnover in research and development. These sectors include: aerospace (22.7 per cent), computers and office machines (17.5), electronics and components (10.4), pharmaceuticals (8.7), scientific instruments (4.8) and electrical machinery (4.5).

2. See *The West in Transition*, Report by the Canadian Economic Council, Ottawa, 1984.

3. It suffices here to mention a few examples. Alberta's proven reserves in oil and gas are four times those of the Middle East; however, these reserves are largely in heavy oil and tar sands and the production cost is relatively high – C$17 per barrel – making Alberta (and Canada) a net importer of oil. Saskatchewan possesses the richest uranium deposits in the world and the province alone has 50 per cent of the Canadian supply (which accounts for 12 per cent of the world uranium reserves). Some 40 per cent of the earth's potash reserves can also be found in Saskatchewan, which supplies 25 per cent of the world consumption.

4. This particular situation is well reflected in the trade structure of British Columbia. In 1980, shipments to first destination within the rest of Canada represented only 15 per cent of the total shipments, while the shipments to other countries represented 43 per cent. The structure is just the opposite for the three other western provinces, their shipments within the rest of Canada being two or three times more important than their shipments to other countries.

5. *White Paper. Proposals for an Industrial and Science Strategy for Albertans – 1985 to 1990*, Government of Alberta, July 1984.

6. Surveys made in France, for instance, show that approximately 5 per cent of the active population have clear projects of creating their own business; only one out of ten does it effectively. One may consider that the same figures apply to Western Canada. In doubling this amount that would mean some 30 000 new firms to be created. 10 per cent of these are estimated technology based firms (i.e. 3 000).

7. *Economic Benefits of Promoting Micro-electronics Through an NSERC II Center – A British Columbia Perspective*, Ministry of Universities, Science and Communications, August 1985; and *High Technology Industry and Post-Secondary Education*, Ministry of Advanced Education and Manpower, Government of Saskatchewan, November 1983.

8. *Women's Programs*, Ministry of Labour, British Columbia. The Alberta Government has also recently created an Advisory Council on Women's issues. It is also worthwhile mentioning that, at the November 1986's Premiers' Conference, each provincial Premier tabled an action plan that outlined measures to increase the participation of women in training and retraining programmes.

9. As shown by the response given to the recently cancelled, Federal Government tax-credit programme for research and development companies (SRTC). However no precise statistics are available on the amount of funds which were leveraged (perhaps 1 per cent of GNP), and which have gone effectively to support of R&D.

10. It is, however, worthwhile mentioning initiatives taken by several universities in the fields of medecine or biology to create private organisations in order to exploit technical competences (e.g. drug testing). Universities retain often minority shares of these organisations which may also receive some funding from the public sector.

11. *Partnership for Growth, Corporate-University Co-operation in Canada*, Judith Maxwell and Stephanie Currie, Regina, 1984.

12. The discussion above refers to executives in the "exploitation" of known resources and is the dominant pattern by far for western Canadian executives in resource industries (heavy oil, timber, mining, fisheries, etc.). However, another mind set is that of "exploration" for resources which characterises the work of such businesses as "wildcat" drilling for new oil fields. Executives engaged in such risky ventures had very different approaches to their business and were much more receptive to technology based start-ups. Nevertheless, it is important to point out that this willingness to accept risk emphasizes the kind of risk found in R&D (risk about one's understanding of nature), not the type of risk found in the marketplace (risk about one's understanding of competitors and buyers' behavior).

13. See *Seeds of Renewal: Biotechnology and Canada's Resource Industries*, Report 38, Science Council of Canada, Ottawa, September 1985.

14. The publication of the *Technology Transfer Catalogue* has been an useful step, if broadly diffused and actively followed-up by appropriate contacts.

15. *The People's Report – A Social and Economic Alternative for British Columbia*, commissioned and published by Solidarity Coalition, March 1985.

Annex

GOVERNMENT POLICIES

ALBERTA

The Government of Alberta is the largest single source of funding for research and development in the province.

In 1983/84, C$458 million was spent on research and development in the natural sciences by all sectors and agencies in Alberta. These funds were applied in the following areas: advancement of science, C$71 million; defence, C$15 million; energy and fuels, C$162 million; environmental issues, C$7 million; health, C$65 million; industrial and economic development, C$128 million; and others, C$10 million.

Research and development expenditures of the Government of Alberta in the natural sciences totalled C$126 million in 1983/84. The major provincial funding departments or agencies were the Alberta Oil Sands Technology and Research Authority (C$40.6 million); the Executive Council (C$19.9 million); the Departments of Hospitals and Medical Care (C$19.8 million); Agriculture (C$16.2 million); Energy and Natural Resources (C$10 million); Environment (C$3.9 million); the Environmental Center (C$3.5 million); and others (C$12 million).

The key performers of government-funded research in the natural sciences were industry (C$31.8 million); Alberta Research Council (C$33.3 million); in-house government departments or agencies (C$15.3 million); hospitals and health organisations (C$16.4 million); universities (C$12.6 million); and others (C$5.2 million).

Main policy initiatives in the field of science technology and innovation during the last ten years include the establishment of:

a) *Research programmes*

- The Alberta Oil Sands Technology and Research Authority (AOSTRA), established in 1975, has committed more than C$400 million to research projects, primarily field projects, with industry handled on a share cost or cost recovery basis.
- The Alberta Heritage Foundation for Medical Research created in 1979 and endowed with C$300 million.
- "Farming for the Future", a programme which supports "unsolicited" research proposals from farmers and research workers; budget C$5 million yearly;
- The Biophysical and Health Effects of Acid Depositions, a jointly funded oil/gas industry government programme to cost C$8 million over seven years;
- The Occupational Health and Safety Research and Education Program, to spend C$10 million over seven years;

- The Alberta Foundation for Nursing Research, to spend C$1 million over five years;
- The Crop Research Program, spending C$850 000 per year since 1974;
- "Maintaining our Forests" Program, spending C$875 000 on R&D since 1979.

b) Technological infrastructure

- Alberta Environmental Center, Vegreville (1981); part of its programme is to work with companies to solve environmental problems;
- Assistance to universities to set up subsidiary companies to exploit university research (e.g. Chembiomed Ltd. at the University of Alberta) and to purchase major items of equipment having potential industrial use (e.g. the Cyber 205 "supercomputer" at the University of Calgary);
- Initiation of the Alberta Research Council (ARC) joint venture R&D programme with companies;
- Creation of the Center for Frontier Engineering Research at the University of Alberta in 1984 to undertake studies in the design and erection of steel structures in the Arctic (private sector, institutional and government support). Alberta government's share was C$1.9 million;
- The creation in 1985 of an Electronics Test Center (C$10 million) at the ARC to undertake testing for production-line quality and certification purposes;
- The expansion in 1985 (C$13 million) of the Alberta Micro-electronics Center, to enable it to perform microchip design and fabrication activities for universities and industry;
- The creation in 1985 at the ARC of a computer/telephone/TV based technological information system for the electronics industry;
- The creation in 1984 of the Office of Coal Research and Technology to provide a focus for funding applied R&D on Alberta's coal resources;
- The construction (completed in 1984, cost C$22 million) of a Coal Research Center;
- The construction (completed in 1984, cost C$8.5 million) of a Food Processing Development Center;
- Support of inter-provincial agencies such as the Prairie Agricultural Machinery Institute (about C$1 million per year) and the Veterinary Infectious Diseases Institute (C$200 000 per year), etc.

c) Venture capital programmes and financial support to innovation

Vencap Equities Alberta Ltd. Backed by a C$200 million loan from the Alberta Heritage Savings Trust Fund and by a debenture and common share offering to the public, vencap provides equity-linked financing to entrepreneurs to develop high risk or innovative businesses.

Alta-Can Telecom Inc. A wholly-owned subsidiary of Alberta Government Telephones, this company assists fast-growth new areas of micro-electronics and telecommunications with risk capital, joint venture investments and technological expertise.

Joint Venture Program. The Alberta Research Council initiated this program to undertake research with companies on an equally-shared joint venture basis. Council participation may be to a maximum of C$500 000 annually per firm for five years.

Product Development Program. This programme of Alberta Economic Development provides front-end financial support to Alberta businesses for improving their products.

Small Business Equity Corporations Program: introduced on May 1984 to encourage the provision of private sector capital for small business.

An *ERDA agreement* signed in 1984 with the Federal Government includes the following activities:

- A forestry assistance programme;
- Coordination of agriculture and food production initiatives;
- Coordination of tourism development;
- Review of transport policy and needs.

BRITISH COLUMBIA

The province's performance in R&D has been poor, even according to Canadian standard (0.7 per cent of GDP against 1.2 per cent for the country as a whole in 1984) and the lowest of the four western provinces. Total industrial intramural R&D expenditures amounted to 46 per cent of the total R&D investment of C$365 million in 1984. The major part of the funding originated from the Federal Government (C$157 million).

Funding from the Provincial Government (C$19 million in 1984) also includes financial support through the Science Council of British Columbia (C$4 million in 1984 for applied R&D awards and industrial fellowships), and the Research Council of British Columbia which undertakes contract and government funded research and provides assistance to the industrial sector. In 1984 British Columbia research contract funding was provided by the Government of British Columbia (32 per cent), other Governments (29.6 per cent) and industry (38.4 per cent).

Major policy initiatives taken by the Provincial Government to stimulate innovation include:

a) The establishment of the *Discovery Foundation* of British Columbia, an independant non-profit society to advance industrial applications of scientific and technological research, particularly through the creation of industrial research parks. These discovery parks currently operate adjacent to the British Columbia Institute of Technology, at the Simon Fraser University and at the University of British Columbia. Further parks are planned at the University of Victoria and other appropriate locations. The Foundation also runs the Discovery Enterprise Program. A venture capital programme is designed to inject seed money into high technology enterprises for development of new products and processes. The Foundation also incorporates the British Columbia Innovation Office, which gives advice and assistance to entrepreneurs and academics interested in the commercial development of innovation.

b) *Measures to promote small businesses*:   to help the modernisation and growth of corporations in the field of new higher value added manufacturing, resource processing and new knowledge based industries, an Industrial Incentive Fund (IIF) has been set up, which operates mainly through repayable loans. A specific programme has been launched for small business.

c) *Financial measures to improve the investment climate*:   the special enterprise zone and tax relief aims at attracting qualified new businesses within specific areas in the provinces. Municipal partnership agreements have a somewhat similar purpose. Through tax relief and reduction of investment cost, the local government encourages private sector development and job creation. Low Interest Loans Assistance (LILA) provides loans for the establishment and expansion of British Columbia companies. Total value of loans provided under LILA since its inception is C$42 million (C$11.1 million in 1983/84).

d) In addition, legislation has been enacted to allow individuals and companies to invest in *small business venture capital corporations* (VCC's). Individuals, corporations and other entities are, under certain conditions, eligible for a tax credit equal to 30 per cent of the equity investment made in a VCC. The function of a VCC is to receive investment and invest in new equity of eligible small business within the province. Four sectors of the economy are eligible for VCC investment. These are manufacturing and processing, R&D tourism and aquaculture.

In 1984 the Federal Government and the Provincial Government signed an *ERDA agreement*, defining on a 50/50 basis a C$525 million five-year programme of co-ordinated initiatives for economic development. This agreement includes a Science and Technology Development Subsidiary Agreement (STDSA) which will result in the granting of C$20 million in order to create centres of excellence in computer science, micro-electronics, robotics and applied mathematics, supplement the operation and equipping of a Biomedical Research Center on the Campus of University of British Columbia and encourage technology transfer through the establishment of industry/university liaison offices in the three universities. Under the same framework an Industrial Development Subsidiary Agreement (IDSA) will provide C$125 million financial assistance to companies undertaking large scale industrial projects in various fields including strategic and knowledge based industries and high value added products with export potential. Also the Small Business Incentives Agreement (SBIA) will provide C$50 million in loans to SMEs for manufacturing, processing and advanced technology developments projects. However, within the ERDA framework the largest part of the investment will be devoted to productivity enhancement and renewal of forest resources (C$300 million). C$40 million will also be devoted to agricultural food regional development and C$10 million to mineral development.

MANITOBA

In recent years, the Government has broadened its commitment towards science and technology and increased technology related initiatives in support of its general strategy. The Department of Industry, Trade and Technology has been instrumental in developing these initiatives especially through the Manitoba Research Council (MRC).

Strategic thrusts were given to the science and technology policy in 1979, when a federal-provincial agreement on economic development (Enterprise Manitoba) resulted in the establishment by MRC of the Industrial Technology Center in Winnipeg and the Canadian Food Products Development Center in Portage la Prairie. These centres provide scientific and technical services to Manitoba's industrial community and the food sectors. Typical projects supported include component and machine design and development, prototype fabrication, process development, CAD/CAM system design, plant layout and product testing. The centres have grown rapidly and through fee for service contracts (over 55 per cent of funding) have expanded operations without increased dependance on government subsidies. They are the principal tool for the adaptation of new technologies by small and medium-sized firms.

Plans were made with the Federal Government to establish an Institute for Manufacturing Technology (renamed Science Place Canada). But the future remains uncertain, since most federal support for operating costs were terminated at the end of 1984. (This institute was initially designed to meet the increasing demand for manufacturing science and production related technologies).

In addition five new initiatives have been funded under the Jobs Fund. The focus of this fund shifted in 1983 from short-term job creation with stabilization policy purpose to long-term economic development. These initiatives are:

a) Technology Commercialisation Program

This programme aims at providing assistance to technological entrepreneurs, promoting business development based on existing R&D and providing seed financing.

b) Strategic research and scholarships programmes

The objective is to encourage technological growth by funding applied research and development and graduate scholarships oriented towards the province's resources, needs and potential (the programme does not fund basic research but concentrates on strategic areas of applied research such as advanced manufacturing systems, electronics, agriculture, food processing, and alternate fuels).

c) Workplace Innovation Center

Its role is to identify, develop and promote innovative approaches to address human-related issues encountered with the introduction of technological change into the workplace.

d) Technology Discovery Program

Its purpose is to advance awareness and effective management of technology by means of seminars, publications, conferences, etc., and to disseminate information, and provide consultation and feedback on technological change and the government's actions.

e) Information Technology Program (Info-Tech)

It is a set of initiatives aimed at significantly increasing the application of information technologies in key areas such as education and public and private sector offices. The programme is jointly funded by industry (more than 20 per cent contribution is expected) and Manitoba government departments in co-operation and co-ordination with the University of Manitoba, Crown Corporations and other special interest groups.

At present there are two component programmes:

- The Education Technology Program which intends to support both educational needs and the growth of the educational computer software (course-ware) industry in Manitoba. Phase I involves the establishment of the Info-Tech Center, an educational communications network, curriculum consulting, and centralised equipment purchasing.
- The Office/Public Technology Program will be using the Info-Tech Center as a practical demonstration of the application of information technologies for use by business. It will undertake to create an awareness of office technology potential and to demonstrate to business how it can improve its efficiency, effectiveness and competitive position. The programme will also conduct pilot projects such as the application of Telidon to test and demonstrate the application of innovative information services and create new business opportunities for Manitobans.

Furthermore to remedy the difficulty of obtaining equity financing for a new business, the Province Government established a *Venture Capital Program* under which it provides 35 per cent of the capital to set up venture capital corporation which in turn invest in small independantly run businesses. Government's participation takes the form of non-voting, 7 per cent shares which are dividend free for the first three years. Private investors make all investment and management decisions. Some 24 venture capital corporations were approved at the end of 1984 with total equity capital of C$6.3 million. Investments created or maintained some 550 jobs.

Specific incentives have been also taken in favour of small business and job creation. For example the Youth Business Start Program provides up to C$4 000 to manitobans from 18 to 24 years of age to cover the start-up cost of new business. A concentrated effort is also made to attract entrepreneurial immigrants to Manitoba from selected areas in the Far East and Europe (through seminars, trips abroad, setting up of permanent representations).

An *ERDA signed between the Federal and Manitoban Governments* includes the following set of measures and expenses:

	Federal government	Manitoba government	Total
	C$ million		
1. Economic development planning	1.50	1.50	3.00
2. Mineral development	14.80	9.90	24.70
3. Transportation development	111.61	26.05	137.66
4. Port of Churchill	38.06	55.09	93.15
5. Urban bus development	25.00	25.00	50.00
6. Agricultural food development	23.00	15.30	38.30
7. Forest renewal	13.58	13.58	27.16
8. Communications and cultural enterprises	13.00	8.00	21.00
Total	240.55	154.42	394.97

SASKATCHEWAN

In 1983 the Saskatchewan Government spent C$42.1 million for R&D activities, about 1.3 per cent of total budgetary expenditures. It performs 58 per cent of these activities.

The Department of Science and Technology (DST) is implementing the Provincial Government policy and is co-ordinating the public research activity, most of which is executed within the Departments of Agriculture, Energy and Mines, Health and Park and Renewable Resources. The Department is also responsible for inter-provincial co-ordination of science and technology policies. DST's initiatives to stimulate R&D and innovation include:

a)   The Inventor Services Program

Inventors and innovators can submit their ideas to the Innovation Institute of Oregon (US) and obtain for a small fee (C$100) a commercial evaluation. The establishment of a business plan is provided within the framework of the Inventor Entrepreneur Program.

b)   The Industrial Research Program

Under this programme, assistance is given to encourage R&D and new product activities. This programme is designed to supplement federal IRDP and IRAP up to 25 per cent of eligible cost (Saskatchewan companies are in majority eligible for the lowest level of federal assistance programme).

c)   The Joint High Technology Research Program

Research projects in areas of mutual government and industry interest may warrant up to a 50 per cent government contribution towards costs.

d)   Request for Proposals Program

Under this programme, up to a maximum of one-third of the cost of the proposal (cost often prohibitive for small Saskatchewan firms) preparation may be reimbursed for proposal covering more than C$10 000.

e)   Feasibility Studies Program

Firms may receive a maximum of C$2 500 for both a technical and commercial feasibility study (with a maximum of two applications per year), when accurate assessment of the technical and commercial viability of a product cannot be carried out in-house.

f)   Research Infrastructure Program

This programme is designed to make available services needed to facilitate a viable advanced technology industry. Initiatives rest with the industry but can be accelerated by government support. Funding through the programme is incremental, over and above that which is provided through other sources.

g)   Capital Equipment Program

The Government provides up to 50 per cent of the cost of specified high technology equipment required to perform specific research contract initiated by industry.

h)   Information Transfer Program

For conferences, seminars and similar functions which bring advanced technology experts and information into the province, organisers may negotiate funding to offset expenses.

Up to C$5 million are invested annually in these eight programmes. Funding source is the Heritage Fund.

Another important initiative is the *creation of the Office of University Research at the University of Regina*. This Office will be the primary contact for enquiries about R&D services on the university campus. It will increase awareness on the campus about the research needs of the private sector and serve as a central source of information on research funds and sponsors.

As in the other provinces a *fiscal system* favourable to innovation and small business has been set up. Taxes on manufacturing industry have been reduced (no provincial corporate income tax for small business). In addition prototypes are being exempted from sales tax. Furthermore job creating investment are encouraged through a C$7 500 grant for each permanent job created (to a maximum of 25 per cent of capital cost and for a minimum investment of C$30 000) while investment in venture capital corporations are stimulated through a 30 per cent tax credit.

A Canada/Saskatchewan subordinate *agreement on advanced technology* has been signed in 1985. This agreement, the first of this type, commits the two Governments to C$33.2 million of expenditures over five years. The plan, which aims at speeding up the development of high technology industries in the province, comprises two major programmes:

Innovation climate assistance with:

– C$1.5 million for development analysis studies;
– C$3 million for industry/university collaboration;
– C$4.5 million for the task force and project teams.

Industrial assistance providing a number of direct incentives including:

– Marketing support (C$2 million);
– First use risk reduction (C$6 million);
– Bridging capital assistance (C$6.2 million of repayable contributions);
– Strategic investment assistance (C$4 million of pre-authorised incentives);
– Industrial investment assistance (C$4 million).

FEDERAL GOVERNMENT

1.   **Incentives for technological innovation**

a)   *Industrial and Regional Development Program (IRDP)*: DRIE program provides financial assistance to eligible projects over all phases of a product life cycle including industrial innovation and plant modernisation. The program also funds non-profit centres related to industrial development (see below). Budget: C$110 million in fiscal year 1984-85;

b)   *Industrial Research Assistance Program (IRAP)*: NRC program, designed to increase the commercial capabilities of companies by providing financial support to research workers engaged in approved industrial research projects of high technical merit with prospects of a high return and a sound business plan. Budget: C$48 million;

c)   *Program for Industry/Laboratory Projects (PILP)*: NRC programme, provides funds to Canadian companies that are willing to undertake further work on research results from

83

NRC, government laboratories and university laboratories to determine whether commercialisation would be economic or whether a specific Canadian opportunity exists. Budget: C$29 million;

d) *Defence Industry Productivity Program (DIP):* DRIE programme, designed to assist Canadian companies in becoming competitive in the supply of defence related products to international markets. Financial assistance is available for research and development, the acquisition of capital equipment, source establishment and marketing feasibility studies. Budget: C$130 million;

e) *Source Development Fund (SDF):* Supply and Services Canada programme, designed to harness the procurement programmes of Federal Government departments and agencies to promote industrial, regional and technical development in Canada. This is achieved by stimulating product innovation and encouraging domestic sources of supply. Budget: C$10 million;

f) *Contracting out: Science and Technology (Unsolicited Proposals) Program:* Since 1972, the federal government has decreed that its mission-oriented science and technology requirements be contracted out to the private sector. The contracting out programme enables the government to provide support to sound, unique unsolicited proposals submitted by an individual or organisation from the private sector on their own initiative to satisfy the science and technology requirement of one or more government departments. Budget: C$25 million;

g) *Investment Tax Credit:* The investment tax credit applies to both current and capital expenditures in respect of scientific research made by a taxpayer in Canada;

h) *Scientific Research Tax Credit (SRTC) Program:* A mechanism devoted to corporations carrying on scientific research and development activities and which are unable because of insufficient income to take full tax deductions and credits to outside investors. Investors are able to invest in debt, shares or a royalty interest in such corporations and are entitled to a tax credit (SRTC) equal to one half of the amount of the funds invested. The scheme has been applied for one year and was withdrawn at the end of 1985;

i) *Departmental Programs:* Several federal departments have programmes designed to assist technology development in specific industries or areas. These include: the Industrial Energy Research and Development Program (IERD), Energy from the Forest (ENFOR, Biomass Production and Conversion), Solar Energy Demonstration program, Energy R&D in Agriculture and Food Program (ERDAF), Conservation and Renewable Energy Demonstration Agreements. Funding is generally available up to 50 per cent of projects cost. Budget allocated to those various programmes is estimated at C$50 million in fiscal year 1984-85;

j) *University/Industry Co-operative Programs:* Through its programmes of scholarships and grants in aid of research, the Natural Science and Engineering Research Council (NSERC) provides Research Manpower Awards (C$4.5 million in 1983-84) and Research Grants (C$31.4 million), the latter aiming at supporting transfer of technology into industry and at initiating substantial projects in areas of national concern (e.g. biotechnology, communications, oceans, etc.).

2. Centres established by DRIE and the former Department of Industry, Trade and Commerce

About 30 such centres have been established since 1970. They fall in several categories. *Industrial Research Institutes* (IRI – units within universities to facilitate use by industrial companies of technical expertise of the universities), *Centers of Advanced Technology* (CAT – units established by universities or provincial research organisations to operate R&D facilities in technologies of importance for Canada), *Industrial Research Associations* (IRA – organisations carrying out R&D work related to a specific industrial sector and supported jointly by companies in the sector), *Industrial Innovation Centers* (IIC – units established by universities to assist inventors and entrepreneurs and teach students to operate new companies). Most of these centres have become self sufficient.

Those centres established in the West include:

- The Office of Industrial Research (IRI), University of Manitoba; Established: 1971; Contract Revenue: C$1.80 million (1981-82); Funding: C$3.76 million (1973/74-1980/81).
- The Center for Ocean Engineering (CAT), British Columbia Research Council; Established: 1973; Contract Revenue: C$1.15 million (1982-83); Funding: C$1.32 million (1972-73-1977-78);
- The Canadian Food Products Development Center (CAT), Manitoba Research Council; Established: 1974; Contract Revenue: C$2.93 million (1983); Funding: C$4.83 million (1973/74-1978/79);
- The Canadian Center for Advanced Instrumentation (CAT), Saskatchewan Research Council; Established: 1981; Contract Revenue: C$3.59 million (1982); Funding: C$ one million (1981/82-1986-87).

More recently DRIE has been establishing Micro-electronics Centers. These centres are to help industry in the application of micro-electronics to production processes and products. They are part of the Micro-electronics Support Program (MSP) which provides for DRIE contribution to approved proposals for feasibility studies, projects, and the development of custom micro-circuit designs. Seven Micro-electronics Centers have been recently established and have been funded under the IRDP, at the following universities:

- Dalhousie University/Technical University of Nova Scotia;
- Université de Moncton/University of New Brunswick;
- University of British Columbia/Discovery Parks;
- University of Alberta;
- University of Manitoba;
- University of Sherbrooke;
- University of Toronto.

3. Support to Universities

Through its three granting agencies, the Natural Sciences and Engineering Research Council (NSERC), the Medical Research Council (MRC) and the Social Sciences and Humanities Research council (SSHRC), the Federal Government distributes about $500 million annually to support research, scientific training and equipment purchases among about 40 Canadian universities.

In all, federal direct sponsorship of university research and scientific training represents less than 8 per cent of university operating income. Because most federal funding covers only the incremental research costs, such as supplies, technicians' salaries and equipment, it is estimated that each grant dollar a university receives from Ottawa forces it to spend at least another dollar on facilities, researchers' salaries and other overheads.

4. Federal R&D Infrastructures

Federal R&D laboratories are mostly established in the fields of agriculture, energy, environment and defence. In the financial year 1982-83 the breakdown of their activities in Western Canada is as follows:

	Person/years	*Of which*: professors/scientists	Expenditures (C$ million)
British Columbia	1 673	462	88.4
Alberta	1 308	424	63.9
Saskatchewan	686	181	32.0
Manitoba	1 637	451	95.5
Total West	5 304	1 518	279.8
Total Canada	24 452	7 973	1 327.5

5. Training Assistance

The Canadian Jobs Strategy (CJS) was introduced in June 1985. The strategy, which is backed by multi-year financing provisions, emphasizes private sector co-operation. It comprises six programmes, each targeted towards a specific labour market problem area:

- Skill Investment. This programme is essentially preventive, aimed at workers whose jobs are threatened by technical change or changing market conditions. The Federal Government subsidises training costs and wages for up to three years;
- Job Entry. This programme subsumes the experimental Youth Training Option (YTO) and is aimed at helping unemployed youth who have left school ill prepared for the labour market to acquire work experience and training. It is also aimed at women who have interrupted their careers to work at home, to re-enter the labour force;
- Job Development. A programme aimed at the long-term unemployed. It combines on-site training with an employer and off-site training as required for up to fifty-two weeks. Businesses, community groups, municipalities and individuals can act as employers. The Federal Government subsidises training costs and wages;
- Skill shortages. A programme aimed at alleviating specific skill shortages. Employers are encouraged to develop their own training programmes. Subsidies and contributions towards training costs are available for on-the-job and off-the-job training, for both full and part-time training, for a period of up to three years;
- Innovations. This programme is designed to encourage new ideas, experimentation, innovation, and pilot projects directly related to labour market concerns;
- Community Futures. This programme aims at alleviating the particular problems of communities afflicted by major layoffs or plant closure.

The estimated total of funds allocated among the six programmes of the Canadian Jobs Strategy, including institutional training, during fiscal 1985/86 amounts for C$2.1 billion.

Part II

ACCOUNT OF THE REVIEW MEETING

I. INTRODUCTION

The Committee for Scientific and Technological Policy met in special session in Paris on 16th June, 1987 for the review meeting. The session was chaired by Mr. Green, vice-chairman of the Committee, and attended by all the Examiners except Mr. Michael Proctor, now in Western Australia, who sent a note for this discussion (see Annex).

Mr. Greenwood, Delegate of Canada to the Committee, introduced the Canadian representatives and briefly presented the context of the study. He said that if the Secretariat members were pressed to express their point of view, they would perhaps admit that it was the most difficult review which they had carried out. This was partly due to the fact that they had to deal with four very independent entities. Mr. Greenwood added that among the federal countries, Canada was probably the one where the second level (i.e. provincial) was the most independent from the first (federal) level. This sometimes created very sensitive situations.

Much had happened since the review visit took place. A national science and technology policy had been elaborated. Each of the four western provinces had gone through an election and ministers responsible for science and technology had been replaced. The prices of basic commodities such as oil, potash, and wheat had dramatically decreased, badly affecting the western provinces' economies and leading to budgetary cuts.

These budgetary problems, together with important changes affecting the Saskatchewan Government, had made it impossible for the Science and Technology Policy representative of that province, to attend this meeting.

In conclusion, Mr. Greenwood underlined the important role played in the study by Mr. Vanterpool, Assistant Deputy Minister for Technology, Research and Telecommunications in Alberta. Mr. Vanterpool had been the "moving spirit" of the study, taking the initial steps and further ensuring the co-ordination of the whole process.

In introducing the Canadian representatives of the western provinces *Mr. Vanterpool* expressed the thanks of the four provinces to the CSTP for having agreed on this study. He congratulated the review team for its remarkable capacity of perception during a very short review visit with only two days spent in each province. The provinces had been generally very pleased with the report. Particular thanks needed to be expressed to the Secretariat for its competence and perseverance since the beginning of the study early in 1985. *Mr. Gardner* and *Mr. Reichert* each endorsed the thanks expressed by Mr. Vanterpool and outlined the features in British Columbia and Manitoba that had received specific comment in the report.

The Secretariat underlined the pioneering spirit of Western Canadians, a spirit which is reflected in this review in many respects. It was the first time that OECD had undertaken a review concerning several provinces or regions within a Member country. It was the first time that there was a consolidation in one single document of the Background Report and the Examiners' Report, which are usually separated. This initiative, resulting from a joint production by the Examiners and the Secretariat, had required major efforts in time and resources, and for this reason might not be repeated in the future. Finally it was the first time that a review meeting had not been held in the country concerned since 1969, when the Review of Austria initiated the practice of having the "confrontation" meeting in the Member country being reviewed. The Secretariat noted that while holding the western provinces' review meeting in Paris might, on the one hand, reduce the impact of the study in Western Canada, on the other hand, it could offer the opportunity for a greater and more active participation by delegations from other Member countries.

II. GENERAL REMARKS ON THE REPORT

The report was briefly presented by *Mr. Wolek*, who expressed to the western provinces the warm thanks of the review team for their great hospitality and the spirit of co-operation manifested in the course of the review. The review team was impressed by the pioneering and entrepreneurial spirit of Western Canadians. The human resources were talented and technically competent. The educational infrastructure was good. Capital and finance were also there. In each Provincial Government, professional, competent and dedicated individuals were developing innovative programmes. The report, however, had left open the issue of implementation of these structures. Mr. Wolek felt that several problems deserved being mentioned:

- The problem of reaching "critical mass" in high technology efforts, a problem particularly acute in the western provinces in view of the limited industrial base, and also the reluctance of existing industry to provide necessary support for the development of new industries;
- A second issue was the lack of world class managers; the type of industrialisation that Western Canada had experienced in the past had not contributed to the development of managerial skills needed for successful high technology ventures within a context of fierce international competition; and
- A third major question related to the relationship between levels of governments. There was growing co-operation between the Federal Government and the Provincial Governments as regards policy plans and structures; there were also encouraging achievements in co-operation between the provinces themselves. However, the general situation was characterised by an extremely fierce competition for federal funding. During the review, the team had great difficulties in identifying what the Federal Government wanted from the regional entities, which themselves did not seem able to express their needs clearly.

Against this background, the review team was very interested to know what had been attempted in the western provinces since the review visit had taken place.

In the case of Manitoba, *Mr. Reichert* indicated that there had been no substantial change since 1985. The most significant progress concerned the level of resources allocated to innovation programmes, with greater focus on certain initiatives. The Technology Commercialisation Program, favourably considered by the review team, had benefited from more resources and had expanded. There were now 50 companies in the programme, which had contributed to the direct creation of 300 new jobs. Technology transfer infrastructures at the industry-university interface had been developed and a new Manufacturing Adaptation Program had been established.

Regarding the three specific recommendations put forward in the OECD report, a planning study had been commissioned from a consultant firm for a Quality Assurance Center, consistent with the proposal made by the OECD review team. Special actions undertaken to widen opportunities for subcontractors had included trade shows launched by

Manitoba Telephone System to inform potential subcontractors, and further efforts with government contracts in relation to natural resource projects in the province. The development of a biomedical engineering industry was consistent with the province's "philosophy" and infrastructure; an important computer firm specialising in medical equipment had been recently attracted to the province. Finally, Mr. Reichert emphasized that this OECD study, simply by its execution, had greatly contributed to enhanced co-operation between provincial government agencies.

Mr. Vanterpool indicated that, compared with 1985, the overall budgetary situation of the Alberta Government had deteriorated considerably as a result of the evolution of energy prices. The overall Government budget now amounted to some C$10 billion with a deficit of about C$3 billion. So all departments had suffered cutbacks. A Department of Technology, Research and Telecommunications was established in 1986. This year, about C$30 million would be made available to the Department to support research and innovation in industry and other organisations.

Since the last two years, the Alberta Telecommunications Research Center had been opened. It is a centre completely led and monitored by industry. A Laser Institute, more university-oriented, had also been opened. However, some management problems had been encountered in this as well as in other organisations at the university-industry interface. A US manufacturer of advanced electronic circuits had established a design and production facility in the province. University offices had been opened for technology transfer and the one established at the University of Calgary had been especially active (11 licences in 1986).

There had been much more self-help between high-technology firms. The networking process had been encouraged by the Councils of Advanced Technology established in Calgary and Edmonton. In the latter city, the Advanced Technology Project had also been launched with the Chamber of Commerce, the R&D Park Association and the City Council as sponsors. The OECD study pointed out the lack of venture capital, particularly for small amounts of money (less than C$100 000). A new fund (SPURT) had been created, which should contribute to solving this issue. Moreover, some big established industries had become more interested in investing in small companies involved in the new technologies. The OECD report pointed out the need for better statistics on high technology companies. A contract had been signed with a consultant company by the Department of Technology, Research and Telecommunications and it should soon be possible to develop an appropriate computerised statistical data base.

Mr. Vanterpool stated that more than half of the Department's money would be spent on micro-electronics and biotechnology. The bulk of the remaining funds would be spent on artificial intelligence, expert systems and robotics in which particular competences existed, notably at the University of Calgary. Mr. Vanterpool regretted to say that Alberta had not yet signed a science and technology sub-agreement with the Federal Government. Exchanges with foreign countries had been encouraging. For example, arising from a "sister-provinces" S&T exchange agreement with Heilongjiang of the Peoples Republic of China, a promising technology transfer appeared to be taking place for the production of special mushrooms; however testing with regard to their receptiveness to western tastes still needed to be carried out.

For *Mr. Gardner* from British Columbia, the OECD study had constituted a landmark in relation to inter-provincial co-operation in this new field of innovation policy. The most significant changes in British Columbia since 1985 had included: the establishment of a Science and Technology Board reporting to the Premier whose membership comprised industrialists and academics, the initiation of a deregulation programme by the Provincial Government, and a more systematic approach to federal procurement by means of regular meetings between Federal Government representatives and a newly-created provincial-industry representative council. University offices for liaison with industry had done a remarkable job since their establishment in 1985/86. A number of initiatives had been developed within the context of the C$20 million sub-agreement on science and technology with the Federal Government, now in operation for three years. A proposal for a Biotechnology Research Center was anticipated to lead to construction start-up later in 1987.

The OECD report pointed out the danger of fragmentation in funding research and British Columbia recognised this problem. An Advanced System Institute had been created aimed at co-ordination of the research efforts of the three British Columbian Universities. Discussions are also taking place with Alberta to avoid duplication of efforts in micro-electronics.

Concerning the specific policy suggestions made by OECD, a new micro-electronics service and products company had been formed to provide among other things training facilities. A new Forestry Research Center had been proposed, with construction scheduled to start later in 1987. It had been difficult to make particular progress in some of the specific social aspects of the Examiners' Report, although it was a matter to which the Provincial Government gave particular attention.

III. TECHNOLOGICAL STRATEGIES

Mr. Aubert introduced the subject of technological strategies by detailing the approach suggested by the OECD study. Western Canada should shift from a natural resource-based economy towards a knowledge-intensive economy. However, the two approaches should not be separated. There was a need to build on the comparative advantages of Western Canada with its natural resource and agriculture base, which would be well "fertilized" by the new technologies. Western Canadian technologies would then acquire a unique image in the North American and international markets. Telecommunications was another area which was particularly advanced in Western Canada and offered significant comparative advantages for further development.

The need to reach "critical masses" in technological efforts was another important aspect to be considered in elaborating technological strategies. A factor affecting the capabilities of reaching "critical mass" was the independent spirit characterising the Canadian provinces. This attitude might sometimes lead to excessive fragmentation of efforts.

A third aspect deserving particular attention in the Western Canadian context was the need to overcome problems related to isolation. That factor might have led to an under-appreciation of technical advances made by international competitors. Major exposure measures were thus required. Isolation could also induce excessive efforts for technological self-sufficiency. A better policy would be to develop an internal absorption capacity, appropriate technology transfer and international co-operation.

Mr. Wolek emphasized the fact that the western provinces had become well aware of the need to focus their efforts on the most promising areas, particularly within the present climate when resources had been cut back due to the decline of commodity prices. Mr. Wolek felt it useful to hear more from the provinces on how they had attempted to relate their support for new technologies to their existing strengths. For instance Manitoba had an interesting approach to revitalise its manufacturing base, and Alberta had attempted to relate strong competence in electronics and software to the petroleum industry.

Another aspect deserving particular attention was the issue of the regionalisation of innovation policy and related federal programmes. In-depth quantitative studies in the United States have shown that those regions which had expanded well were those having benefited most from federal procurement, in particular from space and defence markets. In Canada, federal resources for procurement tended to be concentrated in the areas around Ottawa and Toronto. More generally Mr. Wolek asked whether it would not be appropriate to have a more extended decentralisation of federal money in order to respond more efficiently to provincial needs and opportunities.

Mrs. Saumier-Finch from the Federal Government indicated that Canada had recently defined a national science and technology policy including seven topics:

- Increasing the national R&D effort;
- Developing a technological strategy with the INNOVACTION Program, including a special effort on micro-electronics (C$90 billion are being spent for micro-electronics centres located in seven universities);

93

- Reinforcing networks for technological marketing and assistance;
- Strengthening technology transfer capabilities in universities;
- Strengthening scientific research;
- Better management of the socio-cultural impacts of science and technology; and
- Increase the contribution of science and technology to the regional economic development.

This national strategy had been announced in late March, a few days after the OECD report on Western Canada was derestricted. Several provinces had set up working groups on these agreed topics.

As regards procurement policies, *Mr. Reynolds* indicated that the Federal Government had recently established precise quantitative targets for expenses related to defence contracts in the eastern provinces. Such targets had not so far been defined for the western provinces. Moreover, a programme had been recently established by the Department of Industry and the Department of Supply and Services by which subsidies were provided for start-up industries in defence-related activities. Mr. Reynolds also mentioned the arrangements made with the United States Government in defence production markets in which Canadian firms could compete.

Mr. Gardner noted that several companies in British Columbia had recently benefited from federal government contracts; for instance, one company had taken the opportunity to make a breakthrough in supplying Federal Government telecommunications contracts with the installation of an early warning system in the Artic. British Columbia's companies were also expecting to benefit from some 10 per cent of procurement expenditure related to a major Canadian space programme which had recently been announced, and to construct an ice-breaker vessel for the Federal Government.

In response to the Examiners' remarks, *Mr. Reichert* agreed that technological adaptation was a basic theme in Manitoba's strategy. In the Technology Commercialisation Program there was, for instance, a technology transfer component whereby firms were financially supported when attempting to acquire a particular technology from another organisation. A successful case had been the transfer of an electro-magnet device from the University of Manitoba to a company which now produces and commercialises it on the world market. A technology transfer centre had been also established at this university to help companies to deal with professors and scientists.

The province had also systematically attempted to develop a strategy for transferring technologies from elsewhere rather than a technological self-sufficiency approach. Some people had described this strategy of being "the first to be second".

In relation to the discussion of "critical masses" *Mr. Vanterpool* mentioned that there was a growing interest in precompetitive research on a joint basis involving both university teams and companies. For instance, in the field of artificial intelligence three companies had joined together in a specific programme (PRECARN). As regards the openness on international technology trends, Alberta had delegated one person to the federal team for EUREKA projects, and exciting meetings had been held in the Netherlands and Germany. An oil company had recently made a major investment for encouraging electronic technologies, and another oil company had invested several million dollars in a biotechnology firm. These examples demonstrated the involvement of the natural resource industry in new technologies. The Government of Alberta had recently provided major support for a firm visited by the review team, which was at the forefront of bovine embryo transplantation and reproduction processes. A biotechnology fermentation facility had been set up and could be rented by companies which wished to scale up operations.

In the opinion of *Mr. de Haan*, Delegate of the Netherlands, the western provinces could learn from the smaller European countries which had encountered similar problems in defining their technological strategies and in co-ordinating their efforts. However the western provinces were far from Europe, not only in distance, but also mentally. The western provinces could also be more oriented toward the neighbouring states of the US than towards the Atlantic Provinces and Quebec. Mr. de Haan was also particularly interested to know which approaches had been adopted for encouraging small and medium-sized enterprises, which are an important source of innovation in OECD countries. He noted that the farming community seemed better organised than manufacturing industry in Western Canada and he suggested that particular efforts could be made in the agricultural sector.

In response to Mr. de Haan, *Mr. Gardner* indicated that the Premier of British Columbia had lived in the Netherlands until he was 14 years old. Thus British Columbia was very conscious of relationships with Europe. In addition, within the past few months there had been a delegation from the Netherlands' Embassy in Washington, D.C., lead by the S&T attaché, which held extensive discussions in science policy in British Columbia and The Netherlands and on ways to improve co-operation. It was true that British Columbia exports were strongly oriented towards the neighbouring States of the US. Some 60 per cent of micro-electronics production of British Columbia was sold outside the province, the largest share going to the United States.

The Discovery Foundation was a stimulative approach to the question of assisting small and medium-sized firms in British Columbia which, together with the Discovery Park, provided a multi-service environment with government support for financing start-ups in small high technology companies. In total C$13-14 million had been granted over a three-year period. The success rate had been high. The OECD study had criticised high success rates which could imply that government-funded bodies were not sufficiently risk-oriented. But politicians wanted good results. In relation to the comments made by Mr. de Haan, Mr. Gardner pointed out that when the review team visited the province there were only a dozen firms involved in fish farming. There had recently been a spectacular development of these activities in British Columbia resulting in a current industry of over 200 companies. This had stimulated a great demand for knowledge in biotechnology and related techniques.

Mrs. Saumier-Finch wanted to reassure the Delegate from the Netherlands that Canada found itself a multi-cultural country drawing on the experiences of many countries including Japan, China and, of course, Europe. For instance, the Federal Government had encouraged Canadian companies to negociate within the EUREKA initiative. She reminded delegates that Canada was also a member of the European Space Agency as well as the NASA international programme. Some 98 per cent of the Canadian technology was imported, and there was a need to search for the best opportunities.

Mr. Greenwood provided information regarding the technology diffusion mechanisms operating at the federal level. The National Research Council Field Service employed 150 experienced field specialists, of which 30 to 40 were operating in Western Canada. They were operating with minimal bureaucracy and red tape. This activity was closely linked to the Technology Inflow Program of the External Affairs Department, which assists small firms by paying their travel costs related to technology imports.

Mr. Hilsberg, Delegate of Australia, asked whether any forecasting and risk analysis had been made with respect to the price of natural resource commodities. In Australia, 3 per cent of the GNP had been lost in relation to the decline of commodity prices. One might wonder whether this trend was inevitable and permanent, or whether a revival of the commodity prices could be witnessed in the longer term. Commenting on the technology strategies proposed by the report, Mr. Hilsberg felt that it was not clear what was meant by "using the strengths and comparative advantage of the Western Canadian provinces". There was, for instance, no discussion on the added value that the new technologies could bring to the natural resource activities.

Mr. Hilsberg thought also that the report did not sufficiently discuss international aspects bearing on the growth strategy of the western provinces. In particular there was growing competition from countries of the Pacific Rim area and, for instance, Singapore or the Australian State of Victoria were direct competitors with the western Canadian provinces, aiming to attract new technology and related investments.

Mr. Greenwood indicated that, in order to encourage foreign investment, the Federal Government had recently placed investment development specialists, largely hired from the private sector, in eight Canadian embassies to attract foreign investment to Canada.

In response to the questions raised by Mr. Hilsberg, Mr. Vanterpool indicated that the downward trend in commodity prices was forecast to last at least ten years (in the view of government officials). In relation to the doubts expressed on technological strategies to be pursued by the western provinces there was, in the opinion of Mr. Vanterpool, not much choice. However, with one seventh of the world specialists in seismics living in Alberta, there was potential for developing advanced seismological instrumentation using micro-electronics, advanced methods for ore identification using artificial intelligence, etc. This example illustrated how natural resource sectors and the knowledge-intensive sectors could be blended.

Mr. Vuorikari insisted on the need to articulate a clear vision for steering the overall economic development in the western provinces, recognising that it takes at least 10 years before new fields begin to develop. There was, in particular, a need to train people appropriately and give them marketing attitudes and skills that the natural resource-based economy did not require. There was a need to build up an appropriate technology infrastructure and research centres, etc. For all these reasons, a clear vision should be provided, and that presently elaborated in Western Canada and supported by the OECD report seemed reliable.

Mr. Yoshikawa, Delegate from Japan, suggested that the debate between high technology and low technology should be approached through the objective of motivating people. In many respects, mature technology is more important than high technology, as it is the basis for survival. High technology would be used eventually with a view to ensuring the maintenance of mature technologies. The maintenance of mature technology could be a good motivation for people. Mr. Yoshikawa thought that some scenarios, including concrete projects, could be elaborated and proposed to the people of Western Canada. That might be attractive and more efficient than pursuing dreams of high technologies which are not essential for human development.

Mr. Wolek agreed with the comments made by the Delegate of Japan. The Canadian experience had clearly shown that there is a mutual interest between low technology and high technology, and there were many opportunities for spin-off, as illustrated by oil instrumentation or agriculture software. However their mutual nurturing was somewhat difficult to implement in view of differences in the attitudes and background (mind sets), an issue discussed at length in the report.

The Secretariat drew attention to the note prepared by Mr. Proctor, discussing international aspects which had been underlined by the Delegate of Australia and neglected by the report. Moreover, Mr. Proctor's note included recommendations to make more extensive use of the media to help in shifting from a "grain-based economy" towards a "brain-based economy".

IV. RESOURCES FOR INNOVATION AND GROWTH

Mr. Bonnet introduced the topic of human resources. In all countries, universities adapt to industry's needs with certain difficulties and delays. The same phenomenon could be observed in the western provinces. Thus it was crucial that efforts be made within enterprises, which should have an active on-the-job training policy. This policy would be particularly important in high technology fields, where technical change was particularly rapid. Government incentives could help enterprises in efforts to develop on-the-job training. Mr. Bonnet referred again to the lack of "critical mass" in university programmes. In view of the relatively small population density in the western provinces, each university could not invest massively in a large range of programmes, but should specialise in particular fields. In this context, programmes of student exchanges could be developed at the level of Western Canada, or at a broader level. Business management skills were missing in Western Canada and a particular effort was needed to develop capabilities in technological marketing. Meanwhile, it could be useful to attract foreign specialists who had been exposed to international competition. This required a global marketing policy aimed at selling the overall climate for innovation and growth in the provinces, and making them attractive for business creation. The image of the four provinces needed to be strengthened in the OECD area, and particularly in Europe.

Mr. Bonnet also insisted also on the necessity for better planning of education needs with a view to better-adjusted curricula in the universities. Sometimes the universities were in advance of the labour market. This led to a brain drain and the education investment was thus partly lost. In the course of their visit the Examiners had noted, for instance, that in certain universities half of the engineers specialised in telecommunications were hired away from Western Canada. This brain drain needed to be reduced.

Mr. Vuorikari pointed to several weaknesses in the provision of financial resources, particularly for start-up companies, in Western Canada. The support of natural resource-based industries was still weak, as mentioned above. Governments, particularly at provincial level, were not organised to make efficient use of procurement contracts to stimulate innovation. The support structure for providing financial aid to enterprises and to innovative projects was fragmented; there were dedicated people and organisations, but they were dealing separately with the various aspects (e.g. finance, management, marketing,) which needed to be integrated to support entrepreneurs efficiently. Moreover the number of local specialists having competence in worldwide marketing was rather small. In other OECD countries, and notably in Europe, the trend had been towards the establishment of "single window" mechanisms for financing innovation and entrepreneurship, with networks of people providing all necessary skills to local entrepreneurs. Public authorities should also pay particular attention to those companies which supported high technologies by providing necessary inputs and equipment such as precision instruments or silicon wafers. Such companies were not very interesting for most venture capitalists.

It would also be valuable to fix some target figures for enterprise creation. On the basis of Finland's experience it seemed reasonable to envisage 5 to 10 000 new technology-based start-ups in Western Canada over a 5-10 year period, providing that appropriate support mechanisms were established which would be able to deal with the financial "packaging" of innovative projects.

The Delegate of Germany, *Mr. Borst* commented on the enrolment of students in engineering sciences. The figures provided in Table 14 of the report appeared particularly low compared to the situation in Germany where the rate of enrolment of students in engineering and applied science reached 45 per cent (compared with only 20-30 per cent in the western provinces). Why was the Canadian rate so low? What could be done to increase it? Would it be efficient to reduce the tuition fees? Mr. Borst was in disagreement with the Examiners regarding the forecasting and planning of education needs. Experience had shown the impossibility of forecasting accurately the supply of, and demand for, engineers five or ten years ahead. Thus trying to adjust the curricula to match specific areas of demand was not possible. Rather than attempting to forecast needs, it would be better to provide education curricula as broad as possible in order to facilitate the adaptation process.

In relation to the issue of human and knowledge resources, the Delegate of Sweden, *Mrs. Eliasson*, raised the problem of the regionalisation of research effort and the concentration of research funding. In Sweden a decision had been recently taken to concentrate the research resources in 11 of the 25 higher education institutions. This decision had led to a major debate. Mrs. Eliasson asked about the situation in Canada, and more particularly in Western Canada. She was also interested in the measures taken regarding women and their participation in the economic development, in view of critical comments made by the report on the status of women in Western Canada.

For the Delegate of the European Community Commission, *Mr. Paillon*, the economic picture offered by Canada was, to a certain extent, comparable to that of the EEC. The different regions were unevenly developed and in this respect Mr. Paillon wanted to know whether the regions had made in-depth analyses of what they expected from the Federal Government, and whether the Federal Government had a policy for correcting regional imbalances in scientific infrastructures. Another point needing clarification for Mr. Paillon related to the suggestions put forward in the report regarding financial support. The report said that "the Government may have a vital role to play in filling the gap between reasonable equity and the usual risks carried by the banks, insurance companies, and similar conventional lenders". Were there plans, on the part of the Federal Government, to provide more financial support for innovative projects or firms following, for instance, approaches adopted in Europe?

The Delegate of Finland, *Mr. Koskenlinna*, asked whether, in view of the federal structure of the country, there were difficulties in harmonizing curricula, and in the equivalence of diplomas between universities throughout the different regions.

The Delegate of Italy, *Mr. Caracciolo*, disagreed with a statement made in the report, that the financial status of entrepreneurs and managers was somewhat low in Western Canada compared to other professions, such as lawyers or medical doctors. In general, and by definition, entrepreneurs had a different financial status, since they took risks and behaved differently from other professions. Mr. Caracciolo wanted to know what had been done by governments to meet the needs for developing managerial competence and related training.

The Delegate of Austria, *Mr. Putz*, wanted to know what had been done to follow up the report's recommendations concerning the improvement of industrial relations between employers and employees.

The Delegate of Yugoslavia, *Mr. Matejic*, was particularly interested in actions undertaken to improve managerial skills and more particular in what could be done to avoid repeating business failures?

The Delegate of Spain, *Mr. Cadenas Marin*, raised the question of support provided to the agricultural food industries and, in this context, the question of the development of agricultural extension services which had not responded adequately to needs in many countries. What could be derived from the Canadian experience?

Mr. Hilsberg, Delegate of Australia, felt that the general tone of the report was rather interventionist. There was no consideration in the report for alternative approaches such as reducing goods or inputs prices or improving the tax environment of firms. These approaches were very favoured by individual States in Australia for instance. On the contrary, the Examiners were emphasizing measures aimed at shifting the supply curve, suggesting an increase in the number of entrepreneurs, engineers, firms, etc.

In response to the question put forward by the Delegate of Germany, *Mr. Vanterpool* said that the low rate of enrolment in engineering education was more due to a lack of student places than to a lack of interest on the part of students. Moreover, many students, particularly in business administration, started up their own firm while at the university and this reduced considerably the rate of enrolments in the later years. In recent years some industries had laid off engineers; 2 to 3 000 engineers were unemployed; a number of these had decided to set up their own firms. In response to the problem raised by the Delegate of Sweden regarding the concentration of research funds, Mr. Vanterpool agreed that an inherent conflict might exist between the intention of using research funding for supporting regional development and the necessity of keeping the necessary quality and "critical mass" in research efforts. In respect to the improvement of women's status, he felt that very little progress had been made in Alberta during recent years.

Replying to the questions put forward by the Delegate of CEC, *Mr. Vanterpool*, speaking from Alberta's point of view, indicated that the province was seeking more federal support for research. There were important federal research infrastructures in the province, e.g. in agriculture, energy, forestry, and defence. But the federal support as a whole represented only half as much it should be on the basis of the population of the province. In response to concerns expressed by the Delegate of Finland, Mr. Vanterpool said that Alberta universities accepted university degrees granted in other provinces. More problems existed in transferring students from community colleges or vocational schools to universities. It seemed that direct government approaches had not been overly efficient in management training, but successful initiatives had been taken with government support by local groups in helping people to create their own business. An example was provided by the Young Men's Christian Association (YMCA), which has a "hands-on approach" in helping unemployed people to set up their own business, with an excellent rate of success (9 out of 10). More generally it seemed that the education system in Canada was good for training students in analysis, but not as good for training them in synthesis and action. In relation to questions put forward by the Delegates of Italy and Yugoslavia concerning the status of entrepreneurs in Western Canada, Mr. Vanterpool noted that there was relatively little penalty for failure. It was generally considered as a result of bad luck. This was a very positive feature.

Mr. Reichert said that Manitoba attached a growing attention to women's issues. Women constituted an important part of the workforce, and new entrepreneurs. In relation to the discussion on engineering education, a programme addressed to final-year students of engineering faculties had been launched, with successful entrepreneurs providing models. Another programme had also been operated, whereby enterprises hiring engineering graduates and post-graduates were partly subsidised for related costs. This programme had been quite successful, but it had been withdrawn, because the benefits were considered to be limited to a particular group of people. Finally, in relation to problems discussed by the Delegate of Austria, it was worthwhile mentioning the joint effort made by trade unions, industry and government in the establishment of a workplace innovation centre. The first years of operation of this centre were encouraging. It offered a wide range of services to enterprises introducing new technologies which impact on the workforce.

In response to the questions raised by the Delegate of CEC, *Mr. Reynolds* indicated that there was now a deliberate policy on the part of the Federal Government to establish research institutions outside the federal capital area and to spread them all over Canada, as illustrated, for instance, by the creation of the micro-electronics research centres in seven universities. However, the federal involvement in financing R&D would be reduced in the future, and the private sector would be expected to increase its effort. Nevertheless, there was federal support to cover risk capital needs on the basis of the concept of "incrementality". When a project would not proceed without such support, federal aid could amount to 35 per cent of the project's cost.

Mr. Julien provided details on the policy of matching funds for university research. The Federal Government provides the granting councils with one dollar for each dollar that the universities received from the private sector. Such a policy facilitated the creation and development of centres of excellence. A number of industrial research chairs were to be funded at federal level, which should also facilitate industry-university relationships.

Mr. Kavanagh provided further details on engineering education in Canada. The engineering faculties' share of 10 per cent of total enrolments given in the OECD report for the western provinces was reliable for Canada as a whole, although some universities, such as Waterloo or Technical University of Nova Scotia, which specialise in engineering, might have higher ratios. The picture was even worse when considering post-graduates. Enrolments were far too low because many undergraduates were attracted to industry. This problem had been considered by NSERC (Natural Sciences and Engineering Research Council) which had recently established a programme whereby differential stipends would be offered to post-graduates. Mr. Kavanagh confirmed Mr. Vanterpool's statement that it was not a matter of lack of interest by students but a matter of lack of capacity in the education institutions, which were short of places, advanced equipment, and even adequate teaching personnel. Mr. Kavanagh also agreed with the point of view expressed by the Delegate of Germany regarding the difficulty of forecasting demand and supply of engineers. Moreover there were very significant flows of engineers between the provinces, and a global approach would be needed.

On behalf of the Examiners, *Mr. Wolek* responded to several comments. Firstly, in response to comments made by the Delegate of Sweden, he noted the experience of some universities in the United States which had emphasized the development of their research capacities, following models provided by Stanford and the MIT. This policy had not been very successful. They had lost their graduates who had gone elsewhere. On the other hand, some universities had adopted a more flexible approach, not emphasizing research but rather links with local needs and business communities, and thus had been more successful. Responding to the question put by the Spanish Delegate, Mr. Wolek thought that classical extension services in agriculture had an important role to play, with generalists helping in introducing new technologies in the farming communities. However, when real innovative devices or industries had to be developed, specialised people and organisms with new types of skills needed to be established. In this respect, for instance, the agricultural-business centres set up in Western Canada in liaison with university faculties were encouraging attempts to facilitate technology transfer. In relation to enterprise failures and the comments of the Delegation of Yugoslavia,

Mr. Wolek noted a tendency observed by the review team in certain provinces "to bail out" companies, in order to avoid the political problems that such failures would create. This generally resulted in a need for further bail-outs, and thus was counterproductive.

In response to observations made by the Delegate of Australia, *Mr. Aubert* responded that Provincial Governments had little opportunity to influence factor prices and Canadian output prices, which are fixed on the world market. There was also limited scope for Provincial Governments to use tax policies to promote innovation. Thus, it seemed logical to emphasize the supply side. Moreover, the report made clear that there was a certain conservatism in both industry and the universities when considering new developments in the economy of Western Canada. Under such circumstances, the report advocated more government effort in support of active groups and networks. However, compared with a number of OECD countries, the types of policies suggested in the report were not so interventionist.

V. INTER-PROVINCIAL CO-OPERATION

The topic was introduced by *Mr. Bonnet* who noted that the review team was requested to consider the four provinces as an entity. In this context, co-operation would appear to be particularly relevant. Several areas were identified by the review team. A first area was the exchange of experience in order to compare and eventually transfer successful experiences from one province to another. There were for instance different approaches to incubator spaces in Western Canada at the time of the review visit and the provinces would have benefited from mutual exchanges of information on this matter. Another important area for joint approaches was technological research. The review team suggested a pilot programme modelled on the ESPRIT scheme in Europe, whereby organisms and companies of two different provinces could be supported from a central fund financed by the four provinces. Other suggestions put forward by the review team included the development of a data bank for monitoring new technologies and related markets, a technological exhibition and a Western Canadian Open University.

Mr. Reichert noted that first steps had been taken in inter-provincial co-operation in innovation policy with ministers with responsibility for science and technology meeting together in conjunction with the formalization of a national strategy for technological development. They had begun to formalize such a co-operative approach and had given instructions to their respective administrations for proposing concrete projects. In this respect the suggestions put forward by the OECD study had been most useful. A new meeting between ministers was planned for autumn 1987.

Mr. Gardner, noted that the idea of a western technology exhibition of an itinerant character had been favourably received in several provinces. Existing infrastructures for distant learning, notably with the Open Learning Institute in British Columbia and Alberta, constituted a basis for an Open University, but in this regard co-operation between provinces was still embryonic.

Mr. Paillon, Delegate of CEC, was not sure that a project modelled on the ESPRIT programme would be the most appropriate in the Western Canadian context. Better inspiration could be obtained from another recent CEC programme aimed at stimulating exchanges of researchers between universities, or between universities and firms. This programme was less costly than the ESPRIT programme and was very successful.

The Delegate of Germany, *Mr. Borst*, noted that inter-provincial co-operation seemed less developed and institutionalised in Canada than in Germany and other federal countries. On the basis of the German experience he wondered if some form of institutionalisation would not be beneficial, for instance in relation to the co-ordination of higher education and research. It was an area in which co-ordination at the Länder level in Germany had proved to be most useful. Ministers met every two months for two days. Senior officials met every month. Mr. Borst supported also the Examiners' recommendations regarding the development of an Open University. A small Länder in Germany, Northern Westphalia, had achieved good results in this regard. The "clients" were not the usual students, but people seeking part-time education in order to update their knowledge. This distant learning institution had proved to be an efficient tool for developing skills needed for technological adaptation.

For *Mr. Buchanan*, Delegate of the United Kingdom, the discussion had shown that there was a great commonality of problems between individual regions of OECD countries, despite their geographic remoteness, and their political and economical differences. In this respect he wondered if further measures could be taken to facilitate international co-operation on a bilateral basis or a multilateral basis.

Mr. Greenwood, in response to comment made by the Delegate of Germany, indicated that Canada had a Council of Ministers of Education, which was located in Toronto. A yearly meeting of federal and provincial Science and Technology Ministers had now been established. Concerning questions put forward by the Delegate of the United Kingdom, Mr. Greenwood noted that the participation of individual provinces in bilateral agreements had often resulted in co-operation agreements at the level of provincial research institutes. Good results had been noticeable in co-operation arrangements with German institutes. To illustrate what had been done on a multilateral basis, Mr. Greenwood mentioned the example of Canadian provincial participation in OECD actvities. The CSTP work programme is circulated to the Provincial Governments to get their views. It was by this means that the western provinces had decided to request OECD to review their innovation policies.

On behalf of the Director for Science, Technology and Industry, *Mr. Bell* concluded the session by thanking Mr. Green for his very efficient chairmanship of a meeting in which Delegates of Member countries had participated actively. This participation demonstrated that the problems encountered by western Canadian provinces were in many respects common to a number of OECD countries. Mr. Bell thanked also the Canadian participants who had provided frank and detailed answers to questions put to them, and the members of the review team who had prepared a stimulating report.

EXTRACTS FROM THE NOTE PREPARED BY MR. PROCTOR
FOR THE REVIEW MEETING

On a personal level, it was around fifteen years since I had visited Western Canada prior to taking part in the Examiners' visit. It gave me very great pleasure to participate, and to see in those provinces the encouraging trends towards a knowledge-based economy. Having now returned to live in Australia, the more I see the stronger I realise the similarities between Canada and Australia and specifically between the western provinces and Western Australia. I remain convinced of the vitality and potential of both areas in the coming decades.

A number of the Examiners' suggestions to the western provinces are general in character. In my personal view, steps could be taken in specific areas to reinforce the thrust of initial suggestions.

a) The Pacific Rim concept

The movement to establish and to formalize the concept of the Pacific Rim has received a considerable impulse from the setting up of a recurrent PACRIM conference. This conference, which strives to bring together those nations on the Pacific Rim for mutual discussions on a whole range of political, economic and social issues, was held this year in Perth. In 1988 it will be held in Hong Kong. This forum appears to be suitable for those Canadian provinces that aspire to a "Pacific" orientation to expose their wares to the market.

b) Inter-regional co-operation with other countries

The provinces aim at broadening their contacts in the new technologies to promote their industrial base. There is an obvious value in the provinces, either singly or better as a group, arranging bilateral agreements with other areas of the industrialised world that have experience and strength in the new technologies. Such bilateral agreements would serve as a platform for several proposals put forward by the Examiners. As regards the protocol aspects of foreign relations, parallel with government-to-government agreements, there appears to be a route through individual States in other federal countries. The Federal Republic of Germany and Australia are examples where respectively the Länder of Bayern (Bavaria) and Baden-Wurtemburg, and the States of Western Australia and South Australia appear to be eminently suitable partners for the western Canadian provinces.

c) Development of Technopolises

Small populations are usually unable to generate from their own resources sufficient ideas and capital to support a broad spread of innovative and new technology industry. Many countries have resorted to the method of attracting foreigners who may bring with them these attributes. Special immigration programmes have been launched for this purpose. Insofar as Western Canada has a natural relationship to the Pacific, then this area – described by Toynbee as "the ocean of the future" – is the natural source from which to attract immigrants. A recent report has suggested that Canada has been five times more successful than Australia in attracting business migrants from Asia. There is another thrust that may be applicable to Western Canada. Japan has a particular mechanism for technology promotion, namely the "high tech" city, or "technopolis", that blends technical education, an industrial park and social structures at one site. Recently, investors have shown a willingness to establish these complexes outside Japan in partnership or joint venture with local interests. Such focal points would definitively appear to have a place in the mosaic of Western Canada.

d) The impact of regional centres

The establishment of centres of population, industry or education within a region is a feature of most development policies. However, establishment of such regional centres has counterproductive effects that must also be considered, especially in the context of setting up networks. Regional centres can contribute to the depopulation of their hinterland and reinforce the drift from rural to urban society. This can destroy the networks that are actually desired. Once the networks' components have been destroyed it is very difficult to reverse the process. Long distance servicing of rural communities seems less than satisfactory in practice. In view of the known phenomenon that innovation and subsequent industrialisation are localised, it is important to maintain as high a level of local support services as possible to maximise the benefits that an innovative rural community can bring to a region, a country and of course to itself.

e) Changing "mind sets"

The Examiners report discusses the question of "mind sets", and emphasizes that some of the mind sets observed in Western Canada need to be changed. It is essentially a task for education in the broad sense to change attitudes, and education processes require generations to be effective. Generations are not, unfortunately, available in this instance. A parallel may be found in the measures adopted in some OECD Member countries to combat those other modern scourges: AIDS and drug abuse. A key part in the education campaigns adopted by some governments has been the shock tactic with depiction of painful scenes in advertisements and media presentations. Such a "shock approach" might be useful for our purpose.

LIST OF PARTICIPANTS

Chairman: Dr. R. Green,
Deputy Secretary, Department of Science, Belconnen, Australia

Examiners

Mr. F. Wolek,
Director of Research,
College of Commerce and Finance,
Villanova University,
United States

Mr. V. Vuorikari,
Vice President,
Regional Development Fund,
Kuopio, Finland

Mr. F. Bonnet,
Consultant,
"Silicon Search"
Créteil,
France

Review Team Co-ordinator

Mr. J.E. Aubert
Science and Technology Policy Division
OECD

CANADIAN DELEGATION

Mr. W. Jenkins,
Ambassador, Canadian Permanent
Delegation to the OECD

Mr. W. Greenwood,
Deputy Director,
Science, Technology & Communications Division,
Department of External Affairs,
Ottawa

Mr. A. Vanterpool,
Assistant Deputy Minister,
Planning and Co-ordination,
Technology, Research & Telecommunications,
Alberta

Mr. J. Reichert,
Executive Director,
Technology Division, Department of
Industry, Trade & Technology,
Manitoba

Mrs. S. Saumier-Finch,
Director,
International Relations,
Ministry of State for Science & Technology,
Ottawa

Mr. P. Gardner,
Manager Policy & Program Analysis,
Science and Technology,
Ministry of International Trade,
Science & Investment,
British Columbia

Mr. R. Kavanagh,
Director (Scholarships and Fellowships
Programs),
Natural Sciences and Engineering,
Research Council of Canada,
Ottawa

Mr. A. Reynolds,
Regional Executive Director,
Department of Regional Industrial
Expansion,
Winnipeg

Mr. G. Letourneau,
Assistant Deputy Minister,
Ministry of Higher Education and Science,
Quebec

Mr. G. Julien,
Executive Director,
Natural Sciences and Engineering,
Research Council of Canada,
Ottawa

Ms. A. Pollack,
Canadian Permanent Delegation to the OECD

OTHER SCIENTIFIC AND TECHNOLOGICAL POLICY
COMMITTEE DELEGATES

Mr. T. Hilsberg,
Australia

Mr. G. McAlpine,
Australia

Mr. L. Putz,
Austria

Mr. R. Schurawitzki,
Austria

Mr. A. Stenmans,
Belgium

Mr. D. Spaey,
Belgium

Mr. J. Jacobs,
Belgium

Mr. X. Demoulin,
Belgium

Mr. M. Korst,
Denmark

Mrs. L. Tvede,
Denmark

Mr. M. Lähdeoja,
Finland

Mr. M. Koskenlinna,
Finland

Mr. H. Yoshikawa,
Japan

Mr. H. Nakagawa,
Japan

Mr. M. Nanba,
Japan

Mr. S. Ogura,
Japan

Mr. P. Lenert,
Luxembourg

Mr. J.D. de Haan,
Netherlands

Mr. R.J. Smits,
Netherlands

Mrs. C. Lyche,
Norway

Mr. O. Wiig,
Norway

Mrs. A. M. Faisca,
Portugal

Mr. A. Cadenas Marin,
Spain

Mrs. K. Eliasson,
Sweden

Mr. P.A. Karkkainen,
Finland

Mr. P. Bartoli,
France

Mr. W. Borst,
Germany

Mr. P. Kreyenberg,
Germany

Mr. J. Schlegel,
Germany

Mr. M. Schreiterer,
Germany

Mr. D. Deniozos,
Greece

Mr. T. Higgins,
Ireland

Ms. M. Travers,
Ireland

Mr. V. Ludviksson,
Iceland

Mr. A. Caracciolo,
Italy

Mrs. Oddi Baglioni,
Italy

Mr. H. Yoshimura,
Japan

Mr. A. Larsson,
Sweden

Mr. P. Flubacher,
Switzerland

Mr. H. J. Streuli,
Switzerland

Mr. M. Bara,
Turkey

Mr. K. Gülec,
Turkey

Mr. T. Buchanan,
United Kingdom

Mr. P. de Vos,
United States

Mr. R. Piekarz,
United States

Mr. J. Cohrssen,
United States

Mr. V. Matejic,
Yugoslavia

Mrs. J. Lazeta,
Yugoslavia

Mr. J. Paillon,
CEC

Mr. P. Kerr,
CEC

OECD SECRETARIAT

**Mr. J.D. Bell, Head of
Science and Technology Policy Division**

WHERE TO OBTAIN OECD PUBLICATIONS
OÙ OBTENIR LES PUBLICATIONS DE L'OCDE

ARGENTINA - ARGENTINE
Carlos Hirsch S.R.L.,
Florida 165, 4º Piso,
(Galeria Guemes) 1333 Buenos Aires
Tel. 33.1787.2391 y 30.7122

AUSTRALIA - AUSTRALIE
D.A. Book (Aust.) Pty. Ltd.
11-13 Station Street (P.O. Box 163)
Mitcham, Vic. 3132   Tel. (03) 873 4411

AUSTRIA - AUTRICHE
OECD Publications and Information Centre,
4 Simrockstrasse,
5300 Bonn (Germany)   Tel. (0228) 21.60.45
Gerold & Co., Graben 31, Wien 1 Tel. 52.22.35

BELGIUM - BELGIQUE
Jean de Lannoy,
avenue du Roi 202
B-1060 Bruxelles   Tel. (02) 538.51.69

CANADA
Renouf Publishing Company Ltd/
Éditions Renouf Ltée,
1294 Algoma Road, Ottawa, Ont. K1B 3W8
Tel: (613) 741-4333
Toll Free/Sans Frais:
Ontario, Quebec, Maritimes:
1-800-267-1805
Western Canada, Newfoundland:
1-800-267-1826
Stores/Magasins:
61 rue Sparks St., Ottawa, Ont. K1P 5A6
Tel: (613) 238-8985
211 rue Yonge St., Toronto, Ont. M5B 1M4
Tel: (416) 363-3171

DENMARK - DANEMARK
Munksgaard Export and Subscription Service
35, Nørre Søgade, DK-1370 København K
Tel. +45.1.12.85.70

FINLAND - FINLANDE
Akateeminen Kirjakauppa,
Keskuskatu 1, 00100 Helsinki 10   Tel. 0.12141

FRANCE
OCDE/OECD
Mail Orders/Commandes par correspondance :
2, rue André-Pascal,
75775 Paris Cedex 16
Tel. (1) 45.24.82.00
Bookshop/Librairie : 33, rue Octave-Feuillet
75016 Paris
Tel. (1) 45.24.81.67 or/ou (1) 45.24.81.81
Librairie de l'Université,
12a, rue Nazareth,
13602 Aix-en-Provence   Tel. 42.26.18.08

GERMANY - ALLEMAGNE
OECD Publications and Information Centre,
4 Simrockstrasse,
5300 Bonn   Tel. (0228) 21.60.45

GREECE - GRÈCE
Librairie Kauffmann,
28, rue du Stade, 105 64 Athens Tel. 322.21.60

HONG KONG
Government Information Services,
Publications (Sales) Office,
Information Services Department
No. 1, Battery Path, Central

ICELAND - ISLANDE
Snæbjörn Jónsson & Co., h.f.,
Hafnarstræti 4 & 9,
P.O.B. 1131 – Reykjavik
Tel. 13133/14281/11936

INDIA - INDE
Oxford Book and Stationery Co.,
Scindia House, New Delhi 1 Tel. 331.5896/5308
17 Park St., Calcutta 700016   Tel. 240832

INDONESIA - INDONÉSIE
Pdii-Lipi, P.O. Box 3065/JKT.Jakarta
Tel. 583467

IRELAND - IRLANDE
TDC Publishers - Library Suppliers,
12 North Frederick Street, Dublin 1
Tel. 744835-749677

ITALY - ITALIE
Libreria Commissionaria Sansoni,
Via Lamarmora 45, 50121 Firenze
Tel. 579751/584468
Via Bartolini 29, 20155 Milano   Tel. 365083
Editrice e Libreria Herder,
Piazza Montecitorio 120, 00186 Roma
Tel. 6794628
Libreria Hœpli,
Via Hœpli 5, 20121 Milano   Tel. 865446
Libreria Scientifica
Dott. Lucio de Biasio "Aeiou"
Via Meravigli 16, 20123 Milano   Tel. 807679
Libreria Lattes,
Via Garibaldi 3, 10122 Torino   Tel. 519274
La diffusione delle edizioni OCSE è inoltre
assicurata dalle migliori librerie nelle città più
importanti.

JAPAN - JAPON
OECD Publications and Information Centre,
Landic Akasaka Bldg., 2-3-4 Akasaka,
Minato-ku, Tokyo 107   Tel. 586.2016

KOREA - CORÉE
Kyobo Book Centre Co. Ltd.
P.O.Box: Kwang Hwa Moon 1658,
Seoul   Tel. (REP) 730.78.91

LEBANON - LIBAN
Documenta Scientifica/Redico,
Edison Building, Bliss St.,
P.O.B. 5641, Beirut   Tel. 354429-344425

MALAYSIA - MALAISIE
University of Malaya Co-operative Bookshop
Ltd.,
P.O.Box 1127, Jalan Pantai Baru,
Kuala Lumpur   Tel. 577701/577072

NETHERLANDS - PAYS-BAS
Staatsuitgeverij
Chr. Plantijnstraat, 2 Postbus 20014
2500 EA S-Gravenhage   Tel. 070-789911
Voor bestellingen: Tel. 070-789880

NEW ZEALAND - NOUVELLE-ZÉLANDE
Government Printing Office Bookshops:
Auckland: Retail Bookshop, 25 Rutland Stseet,
Mail Orders, 85 Beach Road
Private Bag C.P.O.
Hamilton: Retail: Ward Street,
Mail Orders, P.O. Box 857
Wellington: Retail, Mulgrave Street, (Head
Office)
Cubacade World Trade Centre,
Mail Orders, Private Bag
Christchurch: Retail, 159 Hereford Street,
Mail Orders, Private Bag
Dunedin: Retail, Princes Street,
Mail Orders, P.O. Box 1104

NORWAY - NORVÈGE
Tanum-Karl Johan
Karl Johans gate 43, Oslo 1
PB 1177 Sentrum, 0107 Oslo 1Tel. (02) 42.93.10

PAKISTAN
Mirza Book Agency
65 Shahrah Quaid-E-Azam, Lahore 3 Tel. 66839

PORTUGAL
Livraria Portugal,
Rua do Carmo 70-74, 1117 Lisboa Codex
Tel. 360582/3

SINGAPORE - SINGAPOUR
Information Publications Pte Ltd
Pei-Fu Industrial Building,
24 New Industrial Road No. 02-06
Singapore 1953   Tel. 2831786, 2831798

SPAIN - ESPAGNE
Mundi-Prensa Libros, S.A.,
Castelló 37, Apartado 1223, Madrid-28001
Tel. 431.33.99
Libreria Bosch, Ronda Universidad 11,
Barcelona 7   Tel. 317.53.08/317.53.58

SWEDEN - SUÈDE
AB CE Fritzes Kungl. Hovbokhandel,
Box 16356, S 103 27 STH,
Regeringsgatan 12,
DS Stockholm   Tel. (08) 23.89.00
Subscription Agency/Abonnements:
Wennergren-Williams AB,
Box 30004, S104 25 Stockholm Tel. (08)54.12.00

SWITZERLAND - SUISSE
OECD Publications and Information Centre,
4 Simrockstrasse,
5300 Bonn (Germany)   Tel. (0228) 21.60.45
Librairie Payot,
6 rue Grenus, 1211 Genève 11
Tel. (022) 31.89.50
United Nations Bookshop/
Librairie des Nations-Unies
Palais des Nations,
1211 – Geneva 10
Tel. 022-34-60-11 (ext. 48 72)

TAIWAN - FORMOSE
Good Faith Worldwide Int'l Co., Ltd.
9th floor, No. 118, Sec.2
Chung Hsiao E. Road
Taipei   Tel. 391.7396/391.7397

THAILAND - THAILANDE
Suksit Siam Co., Ltd.,
1715 Rama IV Rd.,
Samyam Bangkok 5   Tel. 2511630

TURKEY - TURQUIE
Kültur Yayinlari Is-Türk Ltd. Sti.
Atatürk Bulvari No: 191/Kat. 21
Kavaklidere/Ankara   Tel. 25.07.60
Dolmabahce Cad. No: 29
Besiktas/Istanbul   Tel. 160.71.88

UNITED KINGDOM - ROYAUME-UNI
H.M. Stationery Office,
Postal orders only: (01)211-5656
P.O.B. 276, London SW8 5DT
Telephone orders: (01) 622.3316, or
Personal callers:
49 High Holborn, London WC1V 6HB
Branches at: Belfast, Birmingham,
Bristol, Edinburgh, Manchester

UNITED STATES - ÉTATS-UNIS
OECD Publications and Information Centre,
2001 L Street, N.W., Suite 700,
Washington, D.C. 20036 - 4095
Tel. (202) 785.6323

VENEZUELA
Libreria del Este,
Avda F. Miranda 52, Aptdo. 60337,
Edificio Galipan, Caracas 106
Tel. 32.23.01/33.26.04/31.58.38

YUGOSLAVIA - YOUGOSLAVIE
Jugoslovenska Knjiga, Knez Mihajlova 2,
P.O.B. 36, Beograd   Tel. 621.992

Orders and inquiries from countries where
Distributors have not yet been appointed should be
sent to:
OECD, Publications Service, Sales and
Distribution Division, 2, rue André-Pascal, 75775
PARIS CEDEX 16.

Les commandes provenant de pays où l'OCDE n'a
pas encore désigné de distributeur peuvent être
adressées à :
OCDE, Service des Publications. Division des
Ventes et Distribution. 2. rue André-Pascal. 75775
PARIS CEDEX 16.

71055-09-1987

OECD PUBLICATIONS, 2, rue André-Pascal, 75775 PARIS CEDEX 16 - No. 44257 1988
PRINTED IN FRANCE
(92 88 01 1) ISBN 92-64-13056-X